Landowner's Guide to Wildlife Habitat

LANDOWNER'S GUIDE TO
Wildlife Habitat

Forest Management for the New England Region

Richard M. DeGraaf

Mariko Yamasaki

William B. Leak

Anna M. Lester

University of Vermont Press

Burlington, Vermont

PUBLISHED BY UNIVERSITY PRESS OF NEW ENGLAND

HANOVER AND LONDON

Published by University Press of New England,
One Court Street, Lebanon, NH 03766
www.upne.com
© 2005 by University of Vermont
Printed in China

5 4 3 2 1

LIBRARY OF CONGRESS CATALOGING-IN-PUBLICATION DATA
Landowner's guide to wildlife habitat : forest management for the New England region /
Richard M. DeGraaf . . . [et al.].
    p.   cm.
Includes bibliographical references and index.
ISBN 1–58465–467–8 (pbk. : alk. paper)
1. Wildlife habitat improvement—New England. 2. Forest management—New England.
I. DeGraaf, Richard M.
QL84.22.N47L36 2005
639.9'5'0974—dc22   2004022675

CONTENTS

PREFACE

We developed this guide to help New England forest landowners manage their properties as wildlife habitats and as sources of forest products. These goals—improved habitat conditions and timber management—are not incompatible; wildlife habitat improvement, in fact, depends upon timber management. This dependency is a result of historical human impacts on the landscape that have substantially curtailed certain natural disturbance processes that formerly regenerated extensive areas of various early successional habitats needed by a wide variety of wildlife species. While disturbance by wind and pathogens still occurs periodically across the landscape, relatively frequent disturbance by fire, flooding, and beaver do not. Disturbance by fire throughout the extensive oak–hickory forest region of southern New England, by spring flooding that formerly maintained expansive wet meadows along rivers, and by beaver flowages along low-gradient streams formerly created sprawling mosaics of herb/shrub habitats in now developed areas. Such natural disturbances have essentially been eliminated by human activity. Either active management must replace these vibrant, open habitats, or we will witness continued declines in wildlife species that are adapted to postdisturbance environments.

This guide presents options for managing habitats in extensively forested northern New England and in more agricultural/suburban southern New England. It shows for the first time what the various treatments look like on the landscape, so landowners can visualize the results and choose methods that best meet their goals. For all options, habitat changes and wildlife responses are provided. Our goal is to enlist private landowners in the stewardship of New England's forest wildlife. This guide is focused on providing habitat for a diverse array of wildlife species, not on maximizing habitat for any one species.

Although written specifically for New England, this guide is applicable to a much wider area, essentially wherever northern hardwood, red spruce–balsam fir, white pine, or red oak–white pine forests occur. This area includes the Adirondack Mountains of New York, the Lake States and adjacent Canada, and the northern and central Appalachian Mountains in Pennsylvania and West Virginia. The wildlife species involved and the land-use histories of those areas will be somewhat different, but the silvicultural treatments and resulting habitats will be similar to those in New England.

This guide is the third volume in the *New England Wildlife* series. The first, now revised, is subtitled "Habitat, Natural History, and Distribution"; it provides detailed information on the life history, range, and habitats of 338 species of forest wildlife in New England (R. M. DeGraaf and M. Yamasaki. 2001. Hanover, N.H.: University Press of New England. 482 pp.).

The second volume in the series, subtitled "Management of Forested Habitats" (R. M. DeGraaf, M. Yamasaki, W. B. Leak, and J. W. Lanier. 1992. USDA Forest Service General Technical Report NE-144. Radnor, Pa. 271 pp.), presents the silvicultural treatments needed to create or maintain habitat conditions for wildlife diversity in the major forest cover types in New England. It is for professional resource managers.

Whether landowners desire an economic return from forest products or would use such returns to improve wildlife habitat conditions on their forest land, the results are always best realized by following a detailed forest plan. Maintaining a mix of forest conditions for wildlife and producing a sustained yield of valuable wood products require a plan because periodic treatments are required in either case. Plans are best developed with the help of a professional land manager such as a state service forester, consulting forester, or Extension forestry expert. Foresters are licensed in Maine, Connecticut, New Hampshire, and Massachusetts. These specialists will help landowners develop a plan that will meet their goals. Engage qualified loggers to implement the plan. Taking the time to identify qualified logging professionals will enhance the habitat quality, timber values, and appearance of your forest land.

ACKNOWLEDGMENTS

We thank John W. Lanier, habitat biologist, New Hampshire Fish and Game Department; John J. Scanlon, forest project leader, Massachusetts Division of Fisheries and Wildlife; David B. Kittredge, extension forester at the University of Massachusetts; Matthew J. Kelty, professor of forestry, University of Massachusetts; and Paul E. Sendak, research forester, U.S. Forest Service, for their thoughtful comments and suggestions to improve this guide. Mary A. Strong diligently typed the manuscript.

Landowner's Guide to Wildlife Habitat

Introduction

FOREST LANDOWNERS in New England enjoy seeing wildlife on their land. In fact, most surveys of private forest landowners indicate that seeing or enjoying wildlife is a primary reason for ownership. Often, we see only evidence that wildlife has been using the area—tracks, scats, feathers, a dropped antler (fig. 1), or other sign. Have you ever wondered what you could do to create habitat conditions that would attract a greater diversity of species or perhaps benefit some of particular interest? Many landowners would like to improve habitat conditions but are unsure of what to do or how to get started. Others, primarily in southern New England, may associate timber harvesting, especially even-aged management, visually with clearing for suburban development and, therefore, see it as anti-conservation. Such clearing is land conversion, not forest management. This guide is for those who intend to keep their land in forest and to place management options in the context of natural forest disturbances that have historically created an array of wildlife habitat conditions in New England. The net result of either uncertainty or of viewing forest management as land conversion is that owners often do little or nothing to improve habitat conditions, despite their keen interest in wildlife.

Such inaction is unfortunate because most of the forestland in New England is privately owned, and collectively landowners, whether they own 10 or 1,000 acres, could have a great influence on the welfare of many species. Of special concern now are those species that require early successional or young forest conditions. Their futures—whether they thrive or further decline—are largely in the hands of private forest landowners.

Fortunately, since most forest landowners enjoy wildlife, they are willing to improve habitat conditions. This guide is designed to put interest into action and help landowners create specific conditions to benefit the greatest number of forest vertebrates: the amphibians, reptiles, birds,

Fig. 1. Shed antler—always a good find in the woods. Photograph by William M. Healy.

and mammals common to New England woodlands. On small properties, landowners will likely undertake the work themselves. On larger ones, they are encouraged to work with a professional forester to meet their goals. This guide addresses the needs of owners of small and large parcels in northern and southern New England who want to improve wildlife habitat conditions on their lands. Those who do so will have more opportunities to see various species. Many will also have the satisfaction of knowing that they have helped arrest the decline of early successional species by managing their lands to create needed habitats. Wildlife habitat management is entirely compatible with other landowner goals such as aesthetics, privacy, and the production of forest products. As a forest landowner, your goals, your land's extent and condition, and the nature of the surrounding landscape all affect your many opportunities to improve wildlife habitat conditions.

While wood products are typically thought to be the primary products to be derived, many if not most New England landowners value wildlife, recreation, and intangible benefits like privacy and tranquility far more than wood products, and are reluctant to manage their forest at all. This reluctance is often based on the misconception that forest management is somehow incompatible with management for wildlife or recreation. Actually, the reverse is true; increased wildlife diversity (and therefore increased recreational enjoyment) results from a mix of age classes (foresters call them stand sizes), that is, from forest management. Such a desired mix of stand sizes does not result from heedless, unplanned cutting. This so-called "high-grading" — removing the most valuable trees and leaving low quality ones in the remaining stand — creates neither a mix of size classes nor early successional habitats. Rather, it leaves the forest with sparse stands of primarily low-quality trees. In most cases, no specific wildlife conditions exist and the timber value is lost for much of the next century. For example, in southern New England, high-grading is replacing red oak (and its vital mast component) with red maple in many areas. Such practices are not good forest stewardship. Sound management needs to be periodically conducted according to the forest plan to maintain the needed conditions because habitat conditions will change as the forest matures. Some bird species, for example, are present for only a few years as stand conditions change; others are present throughout the life of the stand (table 1).

Forest stewardship requires more than a compilation of individual species habitat associations, however. Management affects species and groups of species differently. Depending on the silvicultural treatment (manipulation of a forest stand by cutting) used, habitats for some species are enhanced; for others, they are degraded. This guide helps landowners strike a balance between creating adequate habitat for declining early successional species and maintaining adequate mature forest habitat for other species that are presently stable or increasing. The reader will note numerous references throughout this guide to maintaining inclusions of certain forest types, maintaining mast-producing trees, snags, coarse woody debris, and other forest attributes that enhance habitat diversity in managed stands by emulating structural patterns that result from natural disturbance processes in unmanaged

Table 1
Effect of Clear-cutting on Breeding, Early Successional Birds (Number of Years)

Bird species	First appear	Become common	Decline
Ruffed grouse (drumming males)	10	15	20
Northern flicker	1	1	7–10
Olive-sided flycatcher	1	1	3–4
Willow flycatcher	1	2	5–7
Tree swallow	1	1	7–10
Winter wren	1	4	7–10
Eastern bluebird	1	1	2
Veery	3	10	20
Swainson's thrush	2	4	15
Cedar waxwing	2	4	7–10
Chestnut-sided warbler	2	4	10
Black-and-white warbler	3	10	[a]
Mourning warbler	2	5	10
Common yellowthroat	2	6	10
Canada warbler	5	15	[a]
White-throated sparrow	1	2	[a]
Rose-breasted grosbeak	3	15	[a]

[a]Present until next cutting cycle. Number of years after clear-cutting an eastern deciduous forest that breeding, early-successional birds first appear, become common, and then decline. We assumed that some residual stems (snags and live trees) remain.
From DeGraaf (1987).

forestlands. Across large landscapes—for example, national forests or state lands—habitats for all species can be provided simultaneously in a shifting mosaic of forest cover types, successional stages, and stand conditions. On smaller ownerships where some timber production is desired, habitat components for groups of species can be provided over time, depending upon the sites involved. On both large and small units, the key is to know the wildlife species that are associated with various forest cover types, timber size classes, and nonforest habitats, and how they respond to silvicultural treatments over time.

This is a "how-to" guide to forest wildlife habitat enhancement for private forest landowners and managers whose goals are primarily concerned with the nontimber values of their land: wildlife enjoyment, privacy, recreation, and aesthetics. It is for people who are interested in producing forest products only to the extent that these other values can be enhanced, and includes:

- New England land use history and forest change.

- Wildlife habitat relationships and the importance of forest structure.

- Methods to recognize habitat improvement opportunities to meet owner goals.

- Procedures to inventory small, nonindustrial ownerships for wildlife habitat conditions/opportunities.

- Depictions of stand conditions to help visualize forest wildlife habitats as they change through time.

- Alternative methods to manage vegetation as wildlife habitats.

- Ways to monitor the effects of habitat improvement.

While this guide focuses on forest habitat diversity, many ownerships contain fields, old orchards, or other nonforest habitats. These habitats contribute to the overall wildlife diversity of the area. Management practices for such habitats are not included here. They range from planting trees, shrubs, and food plots and altering mowing regimes to erecting nest boxes. Please see the list of sources in Further Reading on pages 109–111 if your land contains a fair proportion of nonforest habitats. These guides can be used in addition to the forest management practices outlined here.

Our purpose is to provide information on creating habitat conditions that will support a wide range of species, rather than to make the species present be more easily observed by attracting them to a given site. Such approaches can lead to more problems than enjoyment as bears, raccoons, and other species destroy feeders, gardens, or other property. Our approach focuses on providing habitat to help support wildlife populations, not on attracting individual animals.

The goal of this guide is to show private landowners how to manage their lands for wildlife diversity. To put this goal in perspective, the dynamic nature of the New England landscape is described, including natural disturbance patterns and the effects of past land uses. Of great concern now are the declines of early-successional and young forest species; their conservation needs are not addressed by "letting Nature take its course." Rather, active forest stewardship is needed to maintain a diversity of wildlife on private forestlands. In most cases, forest products and improved timber resource values also result.

Background

Some forest landowners may be reluctant to manage their land with even-aged silvicultural methods, which employ intensive harvests such as clearcutting, for fear of unduly disturbing the site or of fragmenting the forest. Such concerns are certainly appropriate on smaller parcels, say fewer than 25 acres. In the larger forested landscape, however, New England forests are and always have been in a constant state of change. They have always been in the process of changing in response to past glaciation, natural disturbances, introduced pests, and our uses of the land.

Our forests have been changing gradually since the end of the last glaciation in response to climate change. Oaks, beech, hickories, maples, birches, and chestnut, among others, extended their distributions northward at varying rates during the past 10,000 years, creating a shifting pattern of forest cover types on the landscape (fig. 2). Five major types of disturbance have altered New England's forests: windthrow, fire, exotic pests and pathogens, agriculture, and logging. In addition, native insects, beavers, ice storms, drought, flood, landslide, and avalanche have caused minor, but occasionally major, disturbances. For example, the spruce budworm, a native insect, periodically damages millions of acres of spruce–fir forest in northern New England and eastern Canada; recorded outbreaks date back to the 1700s. Also, an ice storm in January 1998 caused severe forest canopy damage on more than 12 million acres in Vermont, New Hampshire, and Maine, although it did not open up the forest as would a fire or major hurricane.

WIND. Catastrophic disturbances from wind occur at long intervals of about 1,000 years in parts of northern New England. In most of the region, however, disturbances — sometimes severe — occur at much shorter intervals. Major hurricanes and windstorms occurred several times during the twentieth century; the last severe hurricane occurred in 1938, when several billion board feet of timber were blown down from Rhode Island to central New Hampshire (fig. 3). The effect of that hurricane was great because 200 years of agriculture and subsequent abandonment produced a high proportion of pine, which suffered far greater damage than did hardwoods. Storms of similar magnitude occurred in 1635 and 1815 and so occur, on average, at 150-year intervals. Local windstorms sufficient to cause some windthrow occur about every 14 years in the White Mountains of New Hampshire. Wind has a dramatic effect on forest overstories, especially in pine and other softwoods, where it commonly sets back the successional stage, but has less impact on the overall species composition because of the presence of a shade-tolerant understory.

Fig. 3. In 1938, a massive hurricane swept through the Northeastern states, blowing down trees on thousands of acres. Here is an example of the blowdown damage on the Bartlett Experimental Forest, New Hampshire. Photograph by USDA Forest Service.

FIRE. Intense forest fires generally occur on dry sites at high or low elevations, and less frequently on mid-slopes. Fire-site soils tend to be glacial outwash sands and gravels, fractured or loose rock, or shallow soils over bedrock (fig. 4). Generally, these sites support mixed-wood or softwood types such as white pine, oak–pine,

Fig. 4. Natural forest fires in the Northeast are not nearly as common or devastating as in the West. However, prior to well-coordinated fire control, severe lightning-caused fires did occur, especially in valley bottoms with sandy soils and upper mountain slopes with shallow, rocky soils. This is the Glen Boulder fire on the side of Mt. Washington (1953). Photograph by USDA Forest Service.

Fig. 2. (foldout) The landscape of southern New England, A.D. 1000 to present. Illustration by Nancy Haver. From R. M. DeGraaf and R. I. Miller. 1996. The importance of disturbance and land-use history in New England: Implications for forested landscapes and wildlife conservation. In **Conservation of faunal diversity in forested landscapes,** edited by R. M. DeGraaf and R. I. Miller. London: Chapman & Hall. © 1996 Chapman & Hall. Used with kind permission of Kluwer Academic Publishers.

pitch pine, or spruce–fir. The impact of fire on forest conditions is more severe than that of wind. Pitch pine barrens occur on repeatedly burned areas that often were originally in white pine or oak–pine. Most bare-rock mountain tops below 3,800 feet in New England are the result of fire, which destroyed organic matter and allowed the thin soils to wash away. Most burns, however, revegetate quickly; for example, the 7,600-acre Beddington Burn in Maine was 97 percent stocked with tree species and other vegetation 5 years after it burned in 1952.

In addition to natural fire, Native Americans in southern New England burned the forest periodically to drive game for hunting, clear fields for planting, and open the forest for traveling. Most such fires had mainly local effects, although some burned until extinguished by rain. Native Americans in southern New England used fire to cultivate the land and open the forest more than those in northern New England. Fire exclusion throughout much of the twentieth century has reduced the occurrence of open habitats in much of New England.

LOGGING. Only a few tracts of land in New England remain unlogged (fig. 5). Much of the logging (as opposed to agricultural land clearing in the 1700s) took place in the mid- to late 1800s when the best softwood trees in mixed-wood stands were cut and softwood stands were clear-cut. Actually, New England's forests have historically been high-graded for timber—"take the best, leave the rest" has been the practice for a long time. Until recently, the only hardwood stands that were heavily cut were those along railroads and those cleared for agriculture in the past. Logging, like windthrow, did not normally affect successional pathways or soils—the same type of forest grew back over time. Intense fires fueled by slash sometimes followed, however, especially in softwood stands on dry sites. Such fires did alter the soil by burning the organic matter, leaving mineral soils that did affect successional

(1000 AD)

Indians annually
burned forests

Indian slash-and-burn
agriculture: sites
cropped 8-30 years

6,000 beaver pelts
shipped from
Springfield, Mass.

1609

(1500)   1524   (1600)   1620   1630   1635

Verrazano explores   Pilgrims land at   Widespread Indian   Major hurricane
New England coast   Plymouth, Mass.   retreats and losses
                                                           from diseases

First closed season
on white-tailed deer
(Rhode Island)

1646

Deer population
very low

early 1800s

(1700)   1750   (1800)   1815   1820–1840

Clearing begins   American   Major   Peak of land clearing
in interior   revolution   hurricane

Great auk extinct

1844

Last turkey killed
in Mass.

1851

Last wolf killed
(New Hampshire)

1860

Elk
exterminated
(Pennsylvania)

1867

1849

California gold rush

1861

Civil War begins

Cougar
extirpated

Last caribou
seen (Maine)

Passenger
pigeon extinct

Heath hen ex

1903

1905

1912

1932

(1900)

1910

1930

Clearing of old-field
white pine

Hardwood seedlings
on cleared sites—
peak of ruffed grouse
populations

Turkey reintroduced

1970

1938

Devastating
hurricane

(1990)

Farm/woodlot/
suburban
landscape

Fig. 5. Signs of past logging activity in the White Mountain region. Photograph by USDA Forest Service.

pathways; for example, much of the paper birch in the White Mountains originated after such fires early in the twentieth century. Today we realize that silviculture involves logging, but the reverse is not true. Logging that leads to erosion, silting of streams and vernal pools, or soil compaction is exploitive logging, not silvicultural logging, which minimizes disturbance to the soil, and maintains the future economic and wildlife habitat values of the developing stand.

More than half of New England was cultivated or grazed in the past, though much of the area has reverted to forest. The effects of agriculture are dramatic: loss of nutrients, changes in soil profiles, and major cover type conversions from hardwoods to old-field white pine farther south.

INTRODUCED INSECTS AND DISEASES. The introduction of exotic insects and diseases has caused rapid and irreversible changes in New England's forests. For example, the range of American chestnut had been expanding slowly across North America for about 8,000 years. Within 50 years of its introduction in 1904, the chestnut blight fungus had eliminated American chestnut as a dominant forest species throughout its range. American elm and American beech have been greatly reduced as canopy trees in much of the region as a result of introduced insects and pathogens.

Insects and diseases have also affected the type of mast (fruit and nuts) available for wildlife by changing the forest composition. Acorns and other hardwood seeds represent the most valuable and energy-rich plant food available in the dormant season, and supported vast flocks of passenger pigeons, once considered the most abundant bird in the world. Oaks increased in importance during the twentieth century as American chestnut declined; oaks have also declined subsequent to the onset of fire protection and defoliation by the gypsy moth, introduced into Massachusetts in 1869. Currently, the hemlock woody adelgid, a small sap-feeding insect native to Japan and China, is spreading northward from southern New England. Minor infestations cause hemlock defoliation and decline; severe infestations kill hemlocks.

AGRICULTURE. Among all sources of forest disturbance, agriculture has had the greatest impact on the forest landscape in New England because it caused major changes in cover types and soils over a wide range of sites (fig. 6). Although fire lowered the productivity of dry-site softwood stands, which reverted to earlier successional stages, such as shrublands, aspen, and birch, fires did not cause major shifts from hardwood to softwood successional paths. Windthrow and logging continue to maintain diversity by initiating earlier successional stages but have little negative impact on the forest ecosystem. Introduced pests and pathogens have had important effects on species composition within forests. The net effect of all these sources of disturbance has been constant change in New England forests.

Pre-European Conditions

Rapid change in the New England landscape began with European settlement. Before settlement, substantial parts of southern New England were quite open due to the presence of native prairies, Native American agricultural clearing and burning, and periodic hurricanes. Throughout the region, abundant beaver meadows and periodic wildfire on dry sites imparted a shifting mosaic of open habitats to the forested landscape as Nature took its course. Beginning about 1,000 to 1,500 years ago, Native Americans south of the Kennebec River in Maine shifted from food gathering to food production and storage, raising corn, beans, squash, pumpkins, and tobacco. Fields were, on average, used for 8 to 10 years until the soil fertility declined and new fields were created. This shifting agriculture, in combination with the presence of native prairies, fire, and windthrow, made for a fairly open landscape in southern New England,

Fig. 6. Cultivation and pasturage in the Northeast, beginning in the early 1600s, caused major changes in cover types and soil conditions. Agricultural areas originally in deciduous forest cover types often grew back to coniferous types, such as white pine and spruce, following farm abandonment. Photograph by William B. Leak.

along the coasts, and along the major rivers of the region. In the forested interior, abundant beaver imparted a substantial component of open habitats in all stages of succession as they created and abandoned flowages.

The resulting abundance of deer, turkey, rabbits, and other game was noted by the earliest English colonists. This abundance of game was produced by Native American agriculture, periodic intentional burning of woodlands, and by hurricanes, which, together, created a patchy landscape of fields and forests in various stages of succession. Native Americans in northern New England did not create such an open landscape; populations were lower, and agriculture was practiced in few, mostly coastal, areas. The net result north of the corn-planting zone (at least 120 frost-free days) was a more heavily forested landscape that was disturbed to a lesser degree by beaver, fire on dry sites, and localized windstorms.

Prior to the Colonial period, much of the northeastern coastal forest of the United States from southernmost Maine to Virginia had a considerable amount and variety of open habitats. The dominant habitat of the region was late-seral or old forest, but extensive grasslands and oak openings were common in eastern North America both along the coast and inland before European settlement. The now-extinct heath hen, an eastern subspecies of the greater prairie chicken, occupied scattered grasslands, native prairies, and blueberry barrens from Virginia north at least to coastal New Hampshire and on the larger offshore islands.

The forest then was quite open in many places. Old forest contained many patches of brushy regeneration and saplings. It was more open and diverse than the economically "mature" 80- to 100-year-old even-aged forest of today. Where agricultural Native Americans had lived, forests reclaimed their old clearings by the early eighteenth century. Native American populations dwindled soon after contact with Europeans. Early settlers who penetrated interior southern New England encountered a landscape that was much more heavily forested than that which had existed a century earlier.

European Settlement Period

Now mostly forested, the New England landscape has undergone dramatic changes over the last 350 years. Land was cleared for agriculture, slowly until the 1750s and then more rapidly until, by the mid-1800s, 75 percent of the arable land in southern and central New England was in pasture and farm crops. One hundred years later, New England was again mostly forested—the result of an era of land abandonment that began soon after the opening of rich farmlands of the Midwest via the Erie Canal in 1825 and the growth of industrial cities and jobs in New England. Except for the miles of stone walls gridding the woods, there is little to indicate that an agrarian society once occupied much of New England.

Around 1910, the white pine that had seeded into impoverished tilled land and dry pastures was cut—the last major land clearing in southern and central New England (fig. 7). Once cut, most sites grew up to hardwoods. Today, about 65 percent of southern New England and more than 90 percent of northern New England is forested. Each year, except on the industrial timberlands in Maine, the age of the forest increases. Much of the current mature forest has been sporadically cut, high-graded, and neglected for most of the past century.

Fig. 7. White pine stands, gridded with old stone walls, are typical of abandoned farmlands in southern and central New England. The understory trees, frequently deciduous species, provide evidence of the original forest type prior to agricultural disturbance. Photograph by Richard M. DeGraaf.

Wildlife Responses to Landscape Change

Wildlife species have undergone dramatic population changes as the New England landscape changed. With settlement, deer and bear populations were reduced to very low levels. Also, large predators such as wolves and mountain lions and furbearers such as beaver were soon extirpated. After the peak of land clearing for agriculture, early successional species including ruffed grouse and vesper sparrows were exceedingly abundant and forest species such as wood thrushes and fishers were absent or quite rare. Extinctions among vertebrate species, despite 350 years of extreme landscape change and early exploitations of wildlife, were surprisingly few. Only three species and one subspecies, all birds, are now extinct: the Labrador duck, great auk, passenger pigeon, and heath hen. Among mammals, only the sea mink and eastern elk, both subspecies, are extinct. Other species, such as rattlesnakes and five-lined skink, are greatly reduced in range. Two conclusions are clear: wildlife species, for the most part, are resilient, and importantly, landscape change exerts more lasting effects than does direct exploitation (fig. 8). A century and a half after the peak of land clearing, the effects of forest regrowth are

Fig. 8. Although early heavy hunting pressure resulted in temporary reductions of certain species, such as snowshoe hares pictured here, the primary cause for long-term changes in wildlife populations has been—and still is—changes in habitat conditions. Photograph provided by Gary A. Getchell.

still occurring, as revealed by declining open-country species and increasing forest species (fig. 9).

Why Can't We Just Let Nature Take Its Course?

Before European settlement, when Nature was free to take its course, habitats for all native wildlife species were continuously being created in a shifting mosaic as natural disturbances occurred and the forest regrew. The landscape in many places was a patchwork of various open habitats, young forest, and old forest. Now, those natural agents of forest disturbance that created grasslands, shrublands, and young forest — to which so many species are adapted — are, for the most part, no longer at work. Native prairies were farmed soon after settlement, wildfire is no longer tolerated, Native American agriculture and burning are no more, and beaver, while abundant, are greatly restricted in their range and activities by stream channel controls.

Disturbances due to wind and pathogens are still at work, but they are very different in scale and frequency from those due to fire, river flooding, and extensive beaver impoundments of presettlement times. These latter disturbances produced a succession of herb-dominated, then shrub-dominated habitats, both more frequently and extensively than wind disturbance.

This is why we can't just "leave it alone and let Nature take its course" and expect all wildlife species to thrive. Doing so would start with a false premise, namely, that we're dealing with a natural forest situation. We're not — 350 years of land use, abuse, and control or elimination of most sources of natural disturbance have resulted in soil changes in most former agricultural areas and an extensive, even-aged mature forest, broken not by burns or beaver flowages but by cities, suburbs, and highways. In order to keep early successional habitats on the landscape for the wildlife species that need them, we need to intentionally and continuously create them.

Even in places where a range of habitats is present, leaving it alone will never keep it as it is. New England landscapes look deceptively stable, especially when we have been familiar with them only for a few years. Some landowners may be lulled into the perception that changing their forest environment through silviculture is somehow working against the grain, and that we're better off to keep the current appearance constant. However, all forest environments in New England are continually adjusting to perturbations caused by climate, ecological processes, and human activities. In the forest ecosystems of New England, the only constant is change, and landowners can

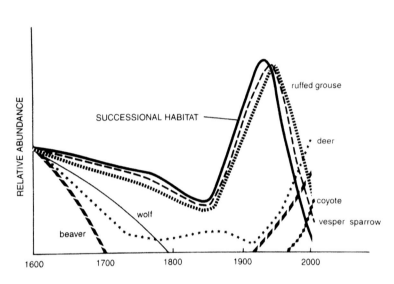

Fig. 9. Schematic depiction of historical changes in representative wildlife species and successional habitat in New England. Several factors are involved in the changes indicated. Wolf, beaver, and deer were persecuted from earliest settlement. Vesper sparrow and ruffed grouse responded to vegetation change with settlement and subsequent land abandonment. Deer and beaver recovered due to protection/reintroduction in the early twentieth century, while coyote colonized the region independently. Note that the abundance of early successional habitat decreased with Native American retreats and declines shortly after initial contact. By Nancy Haver. From Richard M. DeGraaf and Mariko Yamasaki, Fig. 5, p. 15, of **New England Wildlife** © 2001, reprinted with permission of University Press of New England.

guide that inevitable change to achieve wildlife objectives for their land. Small areas may change dramatically due to ice storms or strong winds, but they gradually become mature forest through succession and wildlife diversity declines. Leaving the forest alone—letting Nature take its course—will only allow the landscape to become ever more homogeneous, and as the younger forests age and become similar to the rest of the mature forests in the region, wildlife diversity will continue to decline.

Over the centuries that preceded European settlement, natural disturbances produced a forest substantially more varied both in species composition and in structure in New England, especially southern New England, than we see today. Even if we wait a century or two, the region's forests will never resemble presettlement ones: Native prairies and fire-prone sites are now developed, and floodplain forests cannot occur when floods are prevented by dikes and dams. Time alone will not diversify our forests as much as management will.

For the first time since European settlement, early successional species are declining across eastern North America, and especially in New England. Species that thrive in mature forest, such as squirrels and fisher, are abundant, but most early successional ones are quietly disappearing as the remaining patches of young forest mature. All brushland species are in decline; some, such as the yellow-breasted chat, are already gone, victims of benign neglect. Private forest landowners can readily help reverse the steep declines of others such as whip-poor-wills, eastern towhees, and New England cottontails. The key is to periodically create patches of shrubland and young-forest habitat of sufficient size and proximity to others so that the needed habitats are continually available on the landscape.

New England forests are not tending toward some ideal precolonial condition—there is none. Although there were native prairies in southern New England, natural and Native American-set fires periodically burned large areas, and numerous, extensive beaver flowages imparted an openness to much of the presettlement forest, those conditions did not prevail through the millennia before settlement. Conditions over the past 12,000 years ranged from postglacial steppe, to open spruce forest, and mixed forest, occupied by hunter-gatherers, and then agricultural Native Americans beginning about 1,500 years ago. The New England landscape has always been in a state of change. Now the extent of homogeneous sawtimber-sized forest probably exceeds that which existed at any time previously. We can't turn back the clock. We need to manage the forests that exist now as a result of past human uses of the land, and enhance wildlife habitats in that context. We can reproduce to some extent the historic range of habitat conditions—open upland habitats, regenerating forest, young forests, and old forest—on today's landscape.

No landscape condition is intrinsically ideal or "natural"; we shape the land, and our actions, whether use, abuse, or neglect, dictate which species will thrive and which will decline. In that sense, Nature is what we make it. All species will not survive if we do nothing. To maintain the diversity of New England's wildlife for future generations, active vegetation management is now critically required, not letting Nature take its course.

Understanding Wildlife Habitats

Wildlife Habitat

WILDLIFE HABITAT is the sum of environmental factors—food, water, cover, and their spatial distribution—that each species needs to survive and reproduce in an area. Wildlife species have specific habitat requirements; their abundance and distribution reflect the quantities and quality of habitat available in a given area.

Wildlife habitat management is the manipulation of vegetation structure to influence wildlife populations. Habitats are typically named for wildlife species—for example, ruffed grouse habitat or deer winter habitat. Terms such as "woodcock covert" or "squirrel woods" imply that, within the overall forest, habitat conditions exist that meet the habitat needs for that species. Wildlife biologists use "habitat" to mean a set of conditions required or used by a given species or group of species with similar requirements, or a group of species whose various requirements are met in a given, visually identifiable or distinct environment. For example, sapling stands of northern hardwoods or aspen support a diverse community of species, including redstarts, veeries, rose-breasted grosbeaks, red efts, and red-backed voles, and also provide secure habitat for ruffed grouse. Forest conditions can be maximized for one or more species or, as we intend here, to support a wide diversity of species on your land.

Wildlife populations change in response to many factors, including diseases, predation, extremes of weather, exploitation, and, most importantly, habitat conditions. These factors are not mutually exclusive; several may act in concert or compensate for one another. When habitat conditions for a prolific prey species, such as the white-footed mouse, improve after logging, its population will increase, but opportunistic predators such as owls and weasels will respond to the mouse abundance with increased predation. Populations of all species are always in flux, responding to sets of factors that enhance survival and those that reduce survival. For migratory species, conditions on the wintering grounds or severe weather or accidents during migration are important factors. But by far, the most important factor is habitat condition. Habitat alterations, whether relatively slow as in forest succession or rapid due to fire, logging, or hurricane, produce changes in species composition that are dramatic compared to effects of other factors affecting wildlife populations.

Each species has unique habitat requirements. To be resident in an area, each must find its needs met within the daily activity zones of individual breeding adults. Otherwise, species occurrence is likely to be transitory, temporary, or seasonal. The factors affecting wildlife habitat—food, water, cover, and spatial relationships—are discussed separately and then collectively to illustrate the wildlife community that develops in response to the interaction of species and habitat.

FOOD. A source of energy for growth, maintenance, and reproduction is essential to each species. The plants eaten by grazing or browsing herbivores, such as white-tailed deer, are categorized as preferred, staple, or emergency. Fruits taken by birds can be similarly categorized by how readily they are taken (fig. 10). Food availability varies seasonally, and there are regional differences in food preferences within species. For example, ruffed grouse in winter are fairly dependent on aspen buds—especially male flower buds—in the Lake States, but consume a wide variety of winter foods in New England, where aspen is much less common. In this section we focus on plant foods.

Some species eat a widely varied diet; others are specialists on a few types of food. Species such as white-tailed deer, blue jays, turkeys, gray squirrels, and bears feed on abundant mast when it is available. Blue jays and gray squirrels cache acorns for future use; deer and bear develop a layer of fat—energy for winter—by gorging on fall mast crops. Migratory song birds feed heavily on late-summer fruits to store energy for migration.

The provision of food resources is an extremely important goal for landowners interested in wildlife. In this guide, we present ways to provide fruits and mast through forest management, not by planting trees and shrubs. Fruits such as wild strawberries, raspberries, viburnum, grapes, and cherries are provided naturally as the forest is managed to provide early successional habitat. During the first few years after opening the stand, raspberries and other low-growing fruits are abundant. Later, shrubs such as viburnum and chokecherries provide fruit as the woody growth shades out the herbaceous plants. Fruit is available for about 30 years in sapling and young polestands until pin cherry dies out. Acorns and beech nuts can be provided by reserving large oak, beech, hickories, and other mast trees. Periodic conifer cone crops from hemlock, white pine, spruce, and fir trees whose crowns are free to grow provide abundant seeds for foraging red-breasted nuthatches, pine siskins, red- and white-winged crossbills, evening grosbeaks, chipmunks, red squirrels, and various mice and voles.

Fig. 10a–d. Fruits available from late spring to late fall: (a) woodland strawberry, (b) blueberry, (c) blackberry, and (d) staghorn sumac, an emergency winter food for many bird species. Photographs: (a) Kenneth R. Dudzik, (b) Lois T. Grady, (c) Richard M. DeGraaf, (d) William M. Healy.

In New England, snow and ice storms commonly render food on the forest floor unavailable. For example, woodlands at high elevations in Massachusetts, Vermont, and New Hampshire are unoccupied by wild turkeys in winter because of deep snows. Not all factors affecting food availability can be manipulated through management, but to the extent possible or practical, wildlife food needs, especially of herbivores, should be considered in wildlife habitat management plans.

WATER. Most wildlife species must consume water daily. Water is readily available in most New England woodlands and is rarely a limiting factor for terrestrial species. Aquatic species, however, are restricted in their distributions because they depend on standing or flowing water for breeding, feeding, or overwintering (fig. 11). Species dependent upon fish-free vernal pool breeding habitat include spotted salamander (fig. 12) and wood frog. Creation of ponds or impoundments for wildlife is not necessary to maintain any species populations in New England's forests, but such water bodies are useful to attract some species so that they can be seen and enjoyed. Beaver are now abundant and have created many impoundments.

COVER. Cover is the protective component within an animal's habitat; it provides shelter from the weather and predators. For a New England cottontail, cover is a dense thicket; for a flying squirrel, a tree cavity; for a Swainson's thrush, a small spruce tree in the forest; for a redback salamander, a rotting log. The cover needs of New England's forest wildlife are diverse. Whatever form it takes, cover provides places for escape, roosting or sleeping, travel or reproduction. That's why a homogeneous forest—one in which the dominant trees, regardless of species, are all about the same size—or a plantation contains so few wildlife species, and a forest composed of patches of different size classes and cover types has so many. Wildlife diversity is a function of habitat diversity.

Landowners who want a diversity of wildlife on their land need to be aware of the variety of cover requirements involved, because management can provide the sites that are needed. If an ownership is large enough, at least several hundred acres, using even-aged management to maintain a shifting mosaic of patches of seedling, sapling, poletimber, and mature stands will largely provide the cover needs of most New England forest species. Such a mixture of size classes in a hardwood forest provides habitat for many more species in a forest than does the same-sized area that is all mature forest. Retention of large cavity trees and dead snags throughout the property will provide nesting and denning sites for many others. About a quarter of the species in New England use cavity trees or downed logs, so maintaining a continuous supply is necessary. It is important to retain large live, but declining, trees when regenerating stands; such trees normally will be available throughout the life of the developing stand. Likewise, retaining inclusions of conifers in hardwood stands, or hardwoods in conifer stands, adds species that would otherwise not be present. For example, it only takes a few large hardwoods in a pine stand to attract red-eyed vireos, and a few large conifers in a hardwood stand to attract blue-headed vireos.

Dense brush or seedling stands are uncommon in most New England woodlands unless even-aged management has been practiced recently. In many woodlands, such cover is absent. Shrubland birds are declining rapidly throughout most of the Northeast; whip-poor-will, yellow-bellied cuckoo, chestnut-sided warbler, mourning warbler, indigo bunting, towhee, and field sparrow are just a few of the species that are in decline because their thicket habitats are in decline. Many resident species are also dependent upon woodland thickets or other brushy patches, including ruffed grouse, red-backed voles, and white-tailed deer.

Cover includes nonforest features, such as fields and other grassy areas and water, which provide foraging, escape, and breeding habitat for many species, as well as cover. The cover requirements of all New England forest wildlife are too numerous to mention here. Many are known and readily identifiable, and are routinely provided in normal silvicultural practices. Awareness of the values of wildlife cover in forest management planning will help you meet wildlife needs.

Fig. 11. **Opposite,** Free-flowing forest streams provide breeding habitat for northern spring and two-lined salamanders. Photograph by Kenneth R. Dudzik.

Fig. 12. **Above,** Spotted salamanders breed in vernal pools, often moving hundreds of yards from the adjacent upland forest during the first warm rain in early spring. Photograph provided by Judy Hubley.

Fig. 13. Comparisons of stand entry periods under sustainable, regulated even-aged management for providing continuous early-successional wildlife habitat versus traditional silviculture in the northeastern United States. By Mariko Yamasaki. Adapted from DeGraaf and Yamasaki (2003).

SPATIAL RELATIONSHIPS. The spatial relationships of these factors and their relative abundance and availability largely determine the wildlife species composition—the community—that can occur in a given woodland. Food, cover, and water are reflected to a large degree by the successional stages of various forest types and other cover types, both terrestrial and aquatic. There is a greater likelihood of meeting more species' requirements (that is, a greater wildlife diversity can be expected) where varied habitat conditions are present than where forest conditions are uniform. In forest wildlife management, the goal is to provide patches with different stand structure using uneven-aged methods or patches in different successional stages using even-aged methods. Normally, the changes are in the distributions of present species and size classes, not changes in forest cover types. Such changes are in proportion to stand area and landscape context, so that the forest is not fragmented by "checkerboarding" the whole area into small patches of different habitats. For example, in a 40-acre parcel in agricultural, southern New England, we would not suggest even-aged management with 5- to 10-acre clear-cuts. Small group cuts placed near other open habitat types would be more appropriate. In contrast, in a 500-acre parcel in the extensive forest

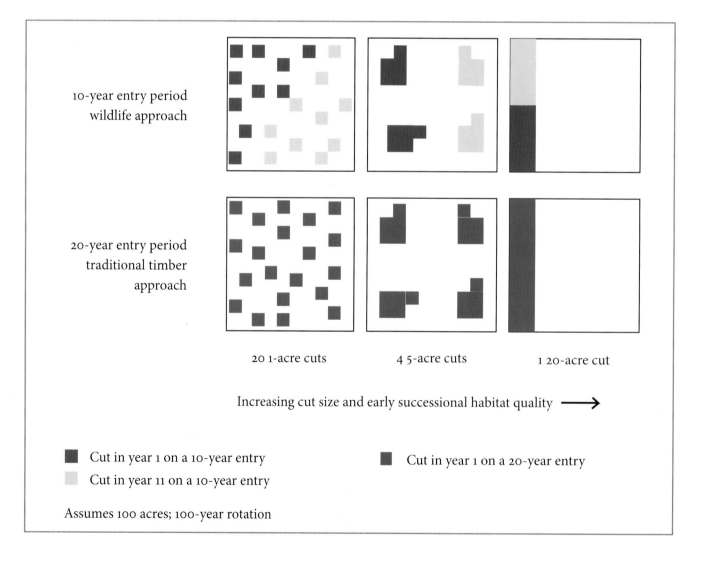

10-year entry period
wildlife approach

20-year entry period
traditional timber
approach

20 1-acre cuts   4 5-acre cuts   1 20-acre cut

Increasing cut size and early successional habitat quality ⟶

■ Cut in year 1 on a 10-year entry   ■ Cut in year 1 on a 20-year entry

□ Cut in year 11 on a 10-year entry

Assumes 100 acres; 100-year rotation

of northern New England, use of clear-cut regeneration would be very appropriate to provide a range of habitat conditions. It is worth noting that many small patches, regardless of parcel size, do not have the same wildlife habitat value as the same acreage in larger patches. Also, clear-cut regeneration for the provision of early successional habitat is generally conducted at about 10-year intervals, rather than the 20-year interval normally used for strictly silvicultural purposes (fig. 13). Clear-cut size would vary with the amount of young forest in the surrounding landscape and size of ownership.

Size of ownership can affect the potential wildlife species that will occupy the site relative to the conditions surrounding the area. Species with small home ranges or territory sizes can be managed easily on smaller units. Portions of the year-round habitat conditions needed by species with large home ranges also can be managed on smaller units depending on the surrounding conditions. For example, if a 40-acre forest parcel is part of an extensive forest, it will be part of the home ranges of species such as great horned owl, goshawk, and bobcat. To the extent that management of the 40-acre parcel improves habitat conditions for prey species such as grouse and cottontails, it is in effect managing habitat for these wide-ranging species. They will use it disproportionately if abundant prey are present. A 40-acre clear-cut, two to three years old, in a northern hardwood forest typically provides breeding habitat for 20 to 30 pairs of willow flycatchers, 2 or 3 pairs of Swainson's thrushes, 1 pair of olive-sided fly-catchers, and many white-footed mice, but only a small part of the home range of a pair of red-tailed hawks that hunt over it and a very small part—about 1 percent—of a moose's home range, which is approximately 4,000 acres.

Community Organization and Structure

Plant communities each support more or less distinct wildlife communities. Here the term "wildlife communities" includes size classes of forest cover types and various upland and aquatic nonforested habitats.

The assemblages of plants, animals, and other organisms occupying the same area comprise a biotic community. These assemblages are mutually sustaining and interdependent. The interacting populations of plants and animals are characterized by continual replacement of individuals, and fluctuate with seasonal and environmental changes.

Food chains, whether simple or complex, link assemblages of herbivores and carnivores to their ultimate plant foods through the process of predation and being eaten. Food webs describe the relationships and interactions between plants (producers), herbivores, insectivores, primary and secondary carnivores, omnivores (all consumers), and decomposers. Within a food web, many plant species normally occur together, and most herbivores and browsers can eat several species. Each species in the food web affects the fluctuations of all other species in some manner (fig. 14). In this way, species present in a community are determined not only by interactions among themselves but also by the adaptations of each species to different environmental conditions.

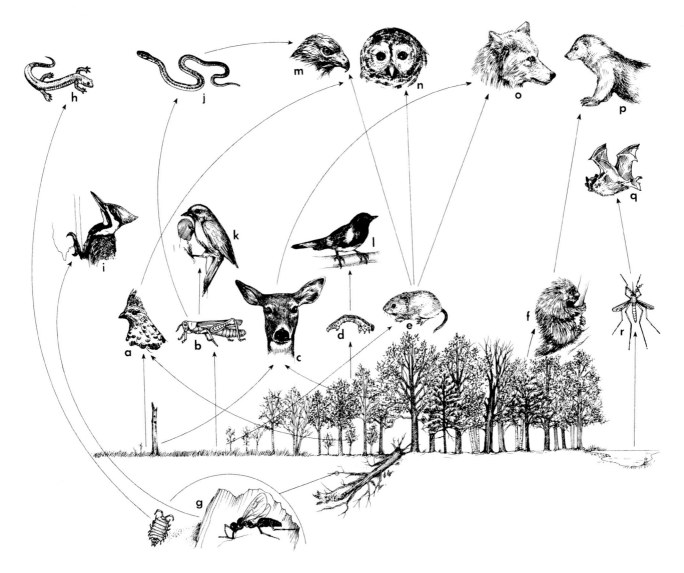

Fig. 14. A representative food web in a typical New England northern hardwood forest sere. Plant parts such as leaves, twigs, buds, and seeds are consumed by herbivores such as ruffed grouse (a), grasshoppers (b), white-tailed deer (c), Bruce spanworm (d), red-backed voles (e), and porcupines (f). Decomposers of plants include sowbugs and carpenter ants (g), which are preyed upon by redback salamanders (h) and pileated woodpeckers (i). Herbivorous insects are consumed by garter snake (j), eastern bluebird (k), and redstart (l), among others. Vertebrate herbivores are consumed by carnivores such as broad-winged hawk (m), barred owl (n), coyote (o), and fisher (p). Bats (q) consume mosquitoes and other flying insects (r). Illustration by Nancy Haver.

Forest habitat management uses methods that predictably relate the major aspects of wildlife species' natural histories (for example, food, water, cover, and spatial requirements) to recognized managed habitats (for example, the various succcessional stages of forest plant communities). These stages contain a unique set of habitat niches—specific arrangements of food, cover, and water—that meet the requirements of particular species.

Habitat niches are, therefore, generally occupied by a set of wildlife species unique to that habitat—species may be either seasonal or year-long habitat occupants. For example, northern hardwood stands can be separated into four habitats based upon the breeding birds that occur in them: (1) regenerating; (2) seedling/sapling; (3) poletimber; and (4) sawtimber, large sawlogs, and uneven-aged stands. The various habitat conditions that come and go, as patch-cuts or clear-cuts are periodically created and new stands grow to maturity, are revealed by the different groups of breeding birds that occur in them over time (fig. 15, pp. 22–23).

The time between successive regeneration cuts needs to be short enough to maintain the presence of early successional bird species in a management area. Otherwise, breeding bird diversity will decline. Mature hardwood stands, whether even-aged sawtimber, old forest, or uneven-aged stands, all have about the same breeding bird composition. At any given time, most of the property will be in medium-aged and mature stands; but the mosaic of regenerating, sapling, poletimber, and sawtimber hardwood stands will provide habitat for about two to three times as many breeding bird species as would occur if the whole unit was mature forest.

Management Practices

Increasing wildlife diversity on your land through creation of a shifting patchwork of habitats over time entails timber harvest. Timber harvests can be as neat as desired. Sometimes there are good reasons to leave slash piles (fig. 16). In general, however, cutting all slash very low, not damaging remaining trees, and making openings fit the landscape go a long way toward reducing the unsightliness of newly cut areas. In areas that were never tilled, irregularly shaped openings can be attractively made following contours. In formerly tilled sites, retaining stone walls in an undamaged condition and fitting cuts to old fields render the opening attractive and somewhat natural: It was once a field, and appears so again for a while. Stone walls define opening sizes that New Englanders have lived with a long time, and are natural boundaries for forest management units in most cases. In New England, newly cut areas "green over" very rapidly compared to other, drier parts of the United States (fig. 17). Within a few years, the area has lost its raw look as various wildflowers, herbs, and forest regeneration appear. Although logging can be temporarily unsightly to some, management provides important early successional habitat that will briefly occupy the site (fig. 18). Seeding skid trails and landings to keep them in grasses and herbs provides habitat for many

Fig. 16. Left, Although the tops and limbs (i.e., slash) left after a logging job can be removed or cut up into smaller pieces (lopped), sometimes the slash is left as is or piled around the borders of an opening to inhibit heavy browsing by deer on the new regeneration. Slash piles also provide habitat for species like long-tailed weasel, deer mouse, and winter wren. Photograph by Richard M. DeGraaf.

Fig. 17. Below, Newly cut areas, such as this 2-year-old clear-cut in northern New England, "green-up" quickly and naturally. Planting is seldom used and rarely needed. Photograph by William B. Leak.

Regeneration

Common yellowthroat
Chestnut-sided warbler
Cedar waxwing
White-throated sparrow
American goldfinch
Willow flycatcher
Mourning warbler
Ruby-throated hummingbird
Philadelphia vireo
Gray catbird

Saplings

Rose-breasted grosbeak
Veery
White-throated sparrow
Black-and-white warbler
Red-eyed vireo
Ovenbird
Canada warbler
Blue jay
Black-capped chickadee
Common yellowthroat

Poletimber

Red-eyed vireo
Ovenbird
Black-throated green warbler
American redstart
Yellow-bellied sapsucker
Veery
Winter wren
Swainson's thrush
Scarlet tanager
Rose-breasted grosbeak

Sawtimber

Ovenbird
Red-eyed vireo
Least flycatcher
Blackburnian warbler
Black-throated blue warbler
Wood thrush
Eastern wood-pewee
Black-throated green warbler
Blue-headed vireo
White-breasted nuthatch

All-aged

Red-eyed vireo
Ovenbird
Black-and-white warbler
Blue-headed vireo
Brown creeper
White-breasted nuthatch
Yellow-bellied sapsucker
Least flycatcher
American robin
Black-throated blue warbler

Old forest

Red-eyed vireo
Ovenbird
American redstart
Rose-breasted grosbeak
Hermit thrush
Yellow-bellied sapsucker
Black-capped chickadee
Eastern wood-pewee
Black-throated green warbler
Scarlet tanager

Fig. 15. Breeding bird composition changes as an even-aged hardwood stand develops over time. Here, the 10 most common species in a New Hampshire northern hardwood forest are listed from most abundant to least. In all cases, the list comprises at least 90 percent of all birds present. Note that the breeding bird composition changes dramatically as the stand progresses from regeneration to saplings to poletimber, and that the bird populations in the sawtimber, uneven-aged, and old forest are not very different. After DeGraaf (1987). Photographs by Richard M. DeGraaf.

Fig. 18a and 18b. Some consider logging slash unsightly, but in New England, it is quite temporary. Smaller pieces decompose within a few years and large pieces provide coarse woody debris used by several species. Shown here is a winter-logged Massachusetts oak–pine stand in the following early summer and the adjacent stand logged 5 years earlier. Photographs by Richard M. DeGraaf.

species—smooth green and garter snakes, brood habitat for grouse and wild turkeys, spring singing grounds for woodcock, and hunting sites for broad-winged hawks, to name a few (fig. 19).

The degree of unsightliness can be controlled by writing specifications regarding slash and postlogging treatment in the logging contract. A service or consulting forester can help you manage your land so it is as attractive as you wish after the job is done.

The delineation of vegetation into forest cover types and size classes, upland and wetland nonforest types, and various types of permanent aquatic

Fig. 19. After a logging job, the roads are often seeded with a mixture of grasses to help stabilize the soil and to provide food (seeds and insects) and cover for a variety of wildlife species, in this case, a ruffed grouse hen and her brood. Photograph by William M. Healy.

habitats provides a basis for determining the habitat niche opportunities or habitat potential available on the site over time.

The distribution or mixing of vegetative types, size classes, and an assortment of features such as cavity trees and thickets largely determines the wildlife communities that occur within a forest. Horizontal and vertical diversity are two components of habitat structure.

HORIZONTAL DIVERSITY. Horizontal diversity or "patchiness" refers to the complexity of the arrangement of plant communities and other habitats (fig. 20). Different forest types or different size classes of any one type have different wildlife communities. The greater the number of forest types or range of size classes present, the greater is the potential that more wild-

Fig. 20. A patchy landscape, such as this area in Maine, where there is a mixture of young stands, old stands, wetlands, water, and so on, is referred to by wildlife biologists as horizontally diverse. Photograph by Mariko Yamasaki.

life species will be present. Still, many wildlife species occur in several forest types and size classes. The horizontal diversity provided by open and wetland habitats contributes greatly to the overall forest wildlife community because these nonforested habitats contain species not normally associated with forest vegetation and provide food, water, and cover opportunities not available in only forested areas.

VERTICAL DIVERSITY. Vertical diversity refers to the extent to which plants are layered within a stand (fig. 21). The degree of layering is determined by the arrangement of growth forms (trees, vines, shrubs, and herbs), by the distribution of different tree species having different heights and crown characteristics, and by trees of different ages of the same species. Stands with a high degree of vertical diversity characteristically develop multiple vegetative layers — overstories with a rich species composition and well-developed herbaceous, shrub understory, and woody midstory layers.

The number of species that occupy a given habitat is, in a large part, a function of habitat structure — its horizontal and vertical structural diversity. Vertical diversity is probably of greatest importance to mature-forest birds and less so to mammals and other species. A mature stand with a well-developed understory and midstory will support more breeding bird species

Fig. 21. Layers of herbaceous vegetation, shrubs, and small, medium-sized, and large trees create vertical diversity within a stand. Photograph by USDA Forest Service.

than one lacking these layers below the canopy. In a forested landscape, horizontal diversity is more important to shrubland and young forest species, and has the greatest impact on overall wildlife diversity.

Other Factors Affecting Wildlife Populations

Foraging times, nesting heights, and soil types are among the factors in addition to habitat structure that affect wildlife occurrence. Factors that permit many bird species to coexist in the same forest include separation of feeding times, for example, hawks feeding during the day and owls at night; separation of breeding times, for example, American goldfinches nesting in late summer, after other finches; and separation of feeding and nesting heights. Factors that commonly influence amphibian and reptile occurrence are soil and moisture regimes, drainage (factors that influence forest type), and cover such as that provided by logs and slash. Small mammals, such as moles, shrews, mice and voles, many of which are seasonally herbivorous, insectivorous, or omnivorous, tend to be habitat generalists, and are not as responsive as birds to changes in forest structure (fig. 22a). Squirrels and other mammals that den in

Fig. 22a–d. Deer mice are ubiquitous (a); flying squirrels need tree cavities (b); and carnivores such as coyote (c) and bobcat (d) need large home ranges with many habitat types. Photographs: (a) Mariko Yamasaki, (b) William M. Healy, (c and d) Thomas J. Maier.

Home Range Legend

Range: 1-2 acres

Chestnut-sided warbler

Mourning warbler

Range: 8-10 acres

Ruffed grouse

Snowshoe hare

Range: 300-500 acres

Sharp-shinned hawk

Barred owl

Range: 4000-6000

Fisher

Moose

Home Range Legend

Range: 1-2 acres

Yellow warbler

Ovenbird

Range: 5-7 acres

Red-bellied woodpecker

Hairy woodpecker

Range: 200-600 acres

Pileated woodpecker

White-tailed deer

Range: 2000-8000 acres

Great horned owl

Black bear

Fig. 23a and 23b. (a) Northern New England. (b) Southern New England. In both northern and southern New England, the larger the ownership, the greater is the number of species whose habitat needs can be met within it. In most cases, species with very large home ranges also use adjacent forestlands. Images by Anna M. Lester.

tree cavities are very much affected by the distribution and structural features of stands that contain both cavity trees and a variety of food sources (fig. 22b). Carnivores and omnivores such as skunks, raccoons, coyotes (fig. 22c), foxes, and bobcats (fig. 22d) need a larger area for feeding than do more localized amphibians, reptiles, and small mammals. White-tailed deer, moose, and black bear can cover large areas, feeding on mixtures of herbaceous, aquatic, and woody vegetation, depending on availability and time of year. Carnivores such as bobcat, lynx, eagles, and other large hawks and owls also cover large areas throughout the year, feeding on various prey depending on availability. Mountain lions and wolves have been replaced in New England by human hunters and coyotes at the top of an intricately intertwined food web.

The size of your ownership and the nature of the surrounding landscape

Fig. 24a and 24b. (a) Large contiguous areas of forested landscape are typical of northern New England; (b) however, much of southern New England is a mixture of forest, field, and residential areas. Photographs by Richard M. DeGraaf.

largely determine which species and how many you can expect to occur. A diverse landscape including ponds, marshes, and fields will contain more species than a more homogeneous one. The larger your ownership, the more likely it will be to contain such habitat elements, and the more likely species with large home ranges will regularly occur because their needs will likely be met. Such species will occur more often as visitors on small ownerships that partly meet their habitat needs. Landowners who want to provide habitat for species with large home ranges will need to look beyond their property boundaries to see whether the adjoining properties contain the habitat features needed by the desired species (fig. 23a and 23b). Wildlife species do not recognize property boundaries unless the boundaries are fenced or lie along abrupt habitat changes, but the larger the ownership, the more diverse the resident wildlife generally will be. This is true in both extensively forested northern New England (fig. 24a) and in partly forested, agricultural, and suburban southern New England (fig. 24b).

Visualizing Forest Change
and Wildlife Responses

IN THIS CHAPTER, we provide visual representations over time of three widely different scenarios for managing a forest: no harvesting, uneven-aged management, and even-aged management. Typical harvesting methods used for uneven-aged and even-aged management are found in Leak et al. (1987). We briefly discuss the effects of these approaches on forest conditions and timber yields, and provide a detailed description of the effects on wildlife habitat conditions. Two hypothetical areas are discussed: (1) a tract of northern hardwood/softwood forest types typical of extensively forested northern New England, and (2) a property typical of the more suburbanized conditions in southern New England supporting oak-pine and hemlock-white pine forest types (fig. 25).

Depending upon your goals, and your forestland's location, size, and the surrounding landscape—whether mostly all forest or somewhat open—you'll find management options that will enhance wildlife habitat diversity. Where the landscape is somewhat agricultural or has extensive wetlands, as is the case in much of southern New England, uneven-aged methods may be most appropriate. Where extensive mature forests prevail, habitat diversity—and therefore wildlife diversity—will increase most dramatically if even-aged methods and clear-cutting are used. You know your goals. We have described various options to indicate how forest conditions will change over time as they are used and the wildlife responses that can be expected as habitats for various species are provided (Appendix A).

Fig. 25. Physiographic map of New England. Northern hardwoods and spruce–fir forests are predominant in northern New England; oak–pine and white pine forests are predominant in southern New England. Map produced by USDA Forest Service. From Keys et al. (1995).

Conditions on this tract were adapted from data collected on the Bartlett Experimental Forest in central New Hampshire, a portion of the U.S. Forest Service's White Mountain National Forest under management by the Northeastern Research Station. This 2,600-acre tract, part of an extensively forested landscape, is divided into three distinct forest types (fig. 26), all about the same size. First there is a lower elevation section (about 700 to 1,000 feet elevation) supporting coniferous stands made up primarily of hemlock and red spruce, mostly 100 to 120 years old with a scattering of white pine and miscellaneous deciduous species— red maple, yellow birch, paper birch, and beech (figs. 27a and 27b). Some of the larger hemlock would be 20 to 30 inches in diameter and perhaps 200 to 300 years old or more (fig. 28). The reason that coniferous species predominate in this area is due to soil conditions—for

Fig. 26. A typical northern New England landscape with a lower section of softwoods (hemlock/red spruce), a midsection of northern hardwoods, and an upper section of steep, rocky land supporting spruce and fir. Images by Anna M. Lester.

example, shallow, wet soils where coniferous species are more competitive than deciduous species.

Just above the coniferous section, where the topography is steeper (at about 1,000-to-2,000-foot elevation), is a tract of deciduous species — the so-called northern hardwood forest type. The primary species are beech, sugar maple, and yellow birch, with smaller numbers of red maple, paper birch, and hemlock (fig. 29). The soils here would be deeper and better drained. Some trees could be over 20 inches in diameter and up to 200 to 300 years old.

Fig. 27a and 27b. (a) Closeup of the low-elevation softwood stand containing mostly spruce with some white pine and hemlock in the background. (b) Hemlock–spruce stands frequently occur on shallow, wet soils; the gray/orange mottling indicates that the soil is saturated with water during most of the year. Photographs by USDA Forest Service.

Fig. 28. Old-growth eastern hemlock stand. Photograph by Richard M. DeGraaf.

The third section is highest on the landscape, between about 2,000 and 3,000 feet in elevation. The topography is steep and the soils are shallow and rocky, supporting spruce and fir and some hemlock with patches of paper and yellow birch. Trees would be smaller and shorter than in the previous two sections, becoming almost scrubby on the mountaintops (fig. 30). This is a reserve area—too steep and rocky for any sort of management activity.

For each of these sections we describe forest conditions and wildlife habitat conditions that would prevail if no management were practiced and those that would develop under uneven-aged and even-aged management. For each forest condition and management option, we give examples of wildlife species that are likely to occur, or not occur, in response to changes in forest structure.

Fig. 29. A typical northern hardwood stand containing beech (gray smooth bark), sugar maple (rough-barked tree on the right), and yellow birch (tree with slight yellowish tinge in the foreground). Photograph by USDA Forest Service.

Fig. 30. High-elevation spruce–fir forest in northern New England. Photograph by USDA Forest Service.

Figure 31 illustrates the effects of even-aged and uneven-aged management on potential habitats available to the amphibians, reptiles, birds, and mammals in northern New England. The fundamental difference between uneven-aged and even-aged managed scenarios is in the sizes of openings that are typically created, and the subsequent structural habitat features that grow, develop, mature, and senesce throughout the life of a stand or forest. Uneven-aged single-tree management and the no-management option produce habitat conditions that are far more similar than different. Table 2 illustrates the types of structural features that silvicultural treatment of forest stands can create over time.

No Management Scenario

Suppose this northern New England tract were simply left alone without any sort of harvesting or other management activity. This situation might be typical of a landowner or agency where the primary interest is recreational or

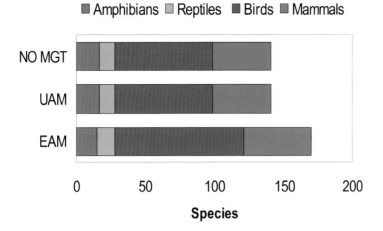

Fig. 31. Potential species numbers by even-aged, uneven-aged, and no management scenarios in New England. By Mariko Yamasaki. After DeGraaf et al. (1992).

scenic — hiking, cross-country skiing, aesthetics, for example. What would happen over time? There would, of course, be no timber revenues from the property. From a distance, the forest tract would continue to look about the same as it did at the beginning (see fig. 26). However, the individual stands (the three sections described earlier) would go through some natural changes — sometimes referred to as natural succession.

CHANGES IN STAND CONDITIONS. The lower elevation stand of coniferous species would increase in the amounts of hemlock and a few spruce — very long-lived species that do not require sunny openings to regenerate and grow. The numbers of pine and deciduous species would tend to decrease. However, from time to time, some of the large, old trees would decay and then die, producing a stand of trees containing an abundance of large, old, hollow coniferous trees and downed woody debris on the forest floor (fig. 32). The forest floor itself would consist of a deep (6 to 12 inches) layer of organic material. Clumps of understory shrubs, such as hobblebush, and herbaceous species, such as partridgeberry, might be common in places (fig. 33). In addition, due to the shallow soils and rooting depths of the coniferous trees, there would be patches of trees, several acres in size, that would be blown down during windstorms. These open patches would quickly regenerate to a mixture of shrubs (e.g., raspberry), noncommercial trees (e.g. pin cherry), and deciduous species such as aspen and paper birch. These occurrences of patch openings would be sporadic and irregular over time.

The northern hardwood stand, just upslope from the coniferous section, would behave much differently under the no-management scenario. Beech and some component of sugar maple would increase over time to the exclusion of most other species. Large, old, hollow beech and sugar maple would be prevalent (fig. 34 and fig. 35) along with hemlock and perhaps a few red oak. Yellow birch would remain in the stand for a long time due to its longevity, but it would not regenerate in any numbers due to its need for sunny openings. In contrast to the just-described coniferous stand, mortality would be

Fig. 32. An old, unmanaged softwood stand showing dead snags, woody material on the ground (coarse woody debris), and some open patches resulting from blowdown. Photograph by USDA Forest Service.

Table 2
Management Scenarios in Northern New England

Within-stand feature	Example wildlife species	No management	Uneven-aged management		Even-aged management		
			Single-tree	Group/patch	Thinning	Shelterwood	Clear-cut
Closed canopy	Barred owl, hairy woodpecker, ruby-crowned kinglet, blackburnian warbler, northern flying squirrel, porcupine	Tree-sized gaps	Tree-sized gaps		X	Canopy closes in time	Canopy closes in time
Partial canopy	Cooper's hawk, northern saw-whet owl, alder flycatcher, great crested flycatcher, bobcat			0.1–0.5 acre gaps	X	When first cut	
Open canopy	Ruffed grouse, northern flicker, eastern bluebird, northern oriole						When first cut
High exposed perches	Osprey, red-tailed hawk, olive-sided flycatcher, common raven			X		X	X
Exposed perches	Eastern phoebe, eastern bluebird, indigo bunting, chipping sparrow			X		X	X
Hardwood inclusions in softwood	Red-eyed vireo, white-breasted nut-hatch, brown creeper, black-throated blue warbler, Baltimore oriole, gray squirrel, white-footed mouse, white-tailed deer	X	X	X	X	X	X
Softwood inclusions in hardwoods	Ruffed grouse, red-breasted nuthatch, golden-crowned and ruby-crowned kinglets, hermit and Swainson's thrushes, northern parula, black-throated green warbler, red squirrel, white-tailed deer	X	X	X	X	X	X
Large cavity trees	Wood duck, common merganser, barred owl, northern long-eared bat, porcupine, raccoon, gray fox, fisher	Abundant	X*	X*	X*	X*	X*
Hard mast	Wood duck, ruffed grouse, wild turkey, blue jay, eastern chipmunk, gray squirrel, northern and southern flying squirrels, deer and white-footed mice, red fox, black bear, white-tailed deer	X	X	Not immediate	X	X	Not immediate
Soft mast	Cedar waxwing, meadow jumping mouse, red fox, black bear, raccoon			X		X	X
Midstory	Acadian flycatcher, veery, wood thrush, solitary vireo, magnolia warbler, American redstart, pine siskin	X	X	Not immediate	X	Not immediate	Not immediate

Within-stand feature	Example wildlife species	No management	Uneven-aged management		Even-aged management		
			Single-tree	Group/patch	Thinning	Shelterwood	Clear-cut
Shrub layer	Alder flycatcher, willow flycatcher, hermit thrush, cedar waxwing, yellow warbler, chestnut-sided warbler, mourning warbler, rose-breasted grosbeak, eastern towhee, dark-eyed junco, snowshoe hare, eastern chipmunk, meadow jumping mouse, black bear, bobcat, white-tailed deer, moose			X		X	X
Herb layer	Garter snake, ruffed grouse, eastern bluebird, golden-winged warbler, indigo bunting, meadow vole, meadow and woodland jumping mice, white-tailed deer			X		X	X
Coarse woody debris	Spotted salamander, red-spotted newt, redback salamander, northern black racer, eastern milk snake, ring-necked snake, masked shrew, eastern chipmunk, long-tailed weasel, black bear, fisher, lynx, bobcat	Abundant	Minimal**	X**	Minimal**	X**	X**

Note: Within-stand features can be provided through various management scenarios in northern New England. Each scenario provides a suite of within-stand features favorable to different wildlife species. This table highlights which within-stand features can be expected under different management scenarios and prescriptions. Note that the wildlife response in large tracts of unmanaged forest over time would be most similar to some form of single-tree uneven-aged management.

*Any timber management activity can reduce the density of large cavity trees. We mark stands to provide a minimum of three to five large live cavity trees per acre in patches and leave strips in addition to all the other uncut large dead trees in the stand. Large cavity trees near water are important structural habitat features and can be easily maintained in riparian buffers.

**Any timber management activity can reduce the recruitment of coarse woody debris on the ground over time. Leaving some large cull trees, particularly hollow ones, in the woods provides denning and shelter features in managed stands.

Adapted from DeGraaf et al. (1992).

Fig. 33. Hobblebush is a common understory shrub in low-elevation softwood stands. Photograph by Richard M. DeGraaf.

tree by tree with few if any large open patches. There would, in other words, be a fairly continuous overstory tree canopy. There would be a scattering of downed woody debris, and a shallow (3 to 4 inches) layer of organic material comprising the forest floor—quite different from conditions under the coniferous stand. The understory would be predominately beech seedlings and saplings with few shrubs or herbs (see fig. 34).

The upper elevation coniferous stand—the so-called reserve area— would behave somewhat similar to the low-elevation coniferous stand. Spruce and fir (with less hemlock) would tend to increase. However, the windthrow patches regenerating to shrubs, pin cherry, and paper birch would be more frequent and larger (fig. 36). There would be some different shrubs—notably mountain-ash, which, with sunlight, produces abundant crops of red berries.

Fig. 34. An old, unmanaged northern hardwood stand with large, decaying trees (beech and sugar maple) and an understory of mostly beech—which would become the predominant species in the stand. In old northern hardwood stands, the trees commonly die and fall over one by one, so there are few open patches that would provide the opportunity for other species to regenerate. Photograph by Kenneth R. Dudzik.

Fig. 35. Fine till soil. Photograph by USDA Forest Service.

WILDLIFE HABITAT CONDITIONS. As shown in table 2, under the "no management" scenario the overstory canopy across the landscape would be fairly closed, and species such as barred owl, hairy woodpecker, Swainson's thrush, and northern flying squirrel, among others, would be present in such conditions. Few large openings (primarily in softwoods) would exist where high exposed perches would provide hunting or song perches for red-tailed hawks or olive-sided flycatchers, or low exposed perches for phoebes or kingbirds to hunt insects in herbaceous openings. Small, shrubby openings and roadside edges would gradually fill in with forest tree species and essentially eliminate the herbaceous and shrubby vegetation, resulting in essentially complete loss of early successional habitat and a loss of habitat diversity over time. Shrub and herbaceous species like the smooth green snake, ruffed grouse, and mourning warbler and the many species that feed on summer fruits and berries would not occur in these conditions.

In the low-elevation coniferous section, the closed canopy conditions would provide habitat for the relatively few but characteristic species associated with old coniferous stands. These include the barred owl, pileated woodpecker, great crested flycatcher, blue-headed vireo, hermit thrush, pine warbler, black-throated green warbler, and species such as the red-breasted nuthatch, crossbills, red squirrel, deer mouse, and others that feed on cone crops. Small gaps due to blow-down would occur infrequently. Unless such disturbances occurred, habitat components such as exposed perches, shrub thickets, shrubs and small trees bearing fleshy fruits (soft mast), and herbaceous growth would be uncommon or absent altogether. Thus, species that hunt from exposed perches, such as red-tailed hawks or phoebes, would likely not be present, nor would shrub-nesting birds like chestnut-sided warbler, or the many birds and mammals that eat fruit in summer.

When a large opening is naturally created through windthrow, a short-lived patch of dense hardwood and softwood regeneration would provide habitat for species such as the ruffed grouse, mourning warbler, white-throated sparrow, towhee, red-backed vole, and snowshoe hare; predators such as ermine and bobcat would likely visit such patches until they fill in with tree species and are no longer thicket habitat. Habitat diversity and wildlife diversity would gradually decline to predisturbance levels as patches of herbs and shrubs disappear. Inclusions of hardwoods in softwood stands would be minimal. If any beech are present, a small crop of nuts would be produced every 3 to 5 years. Very few red oak acorns are produced on softwood sites. Most of the mast produced would come from cone crops. Large cavity trees would be present, and coarse woody debris full of carpenter ants and beetle larvae that black bear feed upon in the springtime.

In the northern hardwood section, the occasional tree that would blow over or die would create only a very small, insignificant canopy gap, allowing only tolerant tree species to regenerate in it. The wildlife community associated with old hardwood stands would not change appreciably in response to these small, sporadic treefall gaps because the overall structure of the stand would not change. Species typically occurring in such a stand include pileated, hairy, and downy woodpeckers, red-eyed vireo, ovenbird, wood thrush, black-throated blue warbler, American robin, and if some spruce or fir is present, black-throated green warbler and solitary vireo. Soft mast (e.g., serviceberry, strawberry, raspberry, pin cherry, and blackberry), and an herbaceous groundcover would not be present. A predominance of beech would periodically produce a substantial hard mast crop every three to five years. There might also be a small, infrequent amount of red oak acorns, making such a stand very important in the fall for black bear, white-tailed deer, and other species. Hard mast diversity would simplify over time with the decline in acorn production. If red spruce or balsam fir are present, cone crops would periodically provide seed crops. Large cavity trees would be present in both hardwood and softwood stands so that large-bodied cavity users like the barred owl, porcupine, and fisher would find nesting and denning here. Large coarse woody debris and down

Fig. 36. High-elevation spruce–fir, showing also a large patch of deciduous trees (mostly paper birch) that regenerated in an area of blowdown. Photograph by USDA Forest Service.

hollow trees would be present for species such as redback salamanders, fisher, and bobcat.

More frequent gap creation in the higher elevation softwood sites would result in stands of predominately spruce and fir, with some paper birch. Typical species include spruce grouse, gray jay, golden and ruby-crowned kinglets, red-breasted nuthatch, hermit thrush, magnolia warbler, blackburnian warbler, white-throated sparrow, long-tailed shrew, red squirrel, and marten. Soft mast (e.g., strawberry, raspberry, pin cherry, blackberry, and mountain-ash) that cedar waxwings and thrushes would feed upon in the summer and an herbaceous groundcover would occur infrequently. Minimal beech or red oak would be expected in these higher elevation sites. Hard mast diversity would probably be less than that on lower elevation sites. Smaller cavity trees would be present; coarse woody debris would be present and might accumulate to a greater degree than on lower elevation sites, and species such as the winter wren could find suitable nesting habitat.

Uneven-Aged Management Scenario

Uneven-aged management, sometimes called selection or partial cutting, is a harvest method whereby trees are removed individually or in small groups (fig. 37). It is a system used by landowners who wish to obtain some timber revenue from their property and would like to keep the landscape somewhat undisturbed, particularly for aesthetic reasons and recreational pursuits.

In this particular scenario, we have assumed that group selection would be used. Every 20 years, there would be a harvest over the entire property except for the high elevation reserved area. Approximately one-fourth of the timber would be removed at each entry. In the coniferous (hemlock–white pine–spruce) section, the groups would average about one-tenth acre in size. About 10 to 15 percent of the coniferous stand would be harvested as groups, and the remainder of the harvest would be tree by tree between groups. The between-group harvest would culture the stand by removing some of the unhealthy and low-quality stems. In the northern hardwood section, the groups would average about one-half acre in size—covering about 10 to 15 percent of the area, and there would be additional tree-by-tree removals between the groups to make up the total harvest of one-fourth of the timber. (Some would prefer to use the term small patches for half-acre groups). A sequence of group selection over 100 years in the northern hardwood stand is illustrated (fig. 38).

Why use groups instead of tree-by-tree removals throughout the stand? There are both good forestry and good wildlife reasons to use groups. In the coniferous stand, the groups are small to favor the regeneration and development of shade-loving hemlock and spruce—but with a component of sun-loving tree species and shrubs. In northern hardwoods, single-tree removals produce understory regeneration composed

Fig. 37. An aerial example of a group selection cutting in northern hardwoods. These groups are like small clearcuts, about ½ acre in size, and round. But various sizes and shapes are possible. Photograph by USDA Forest Service.

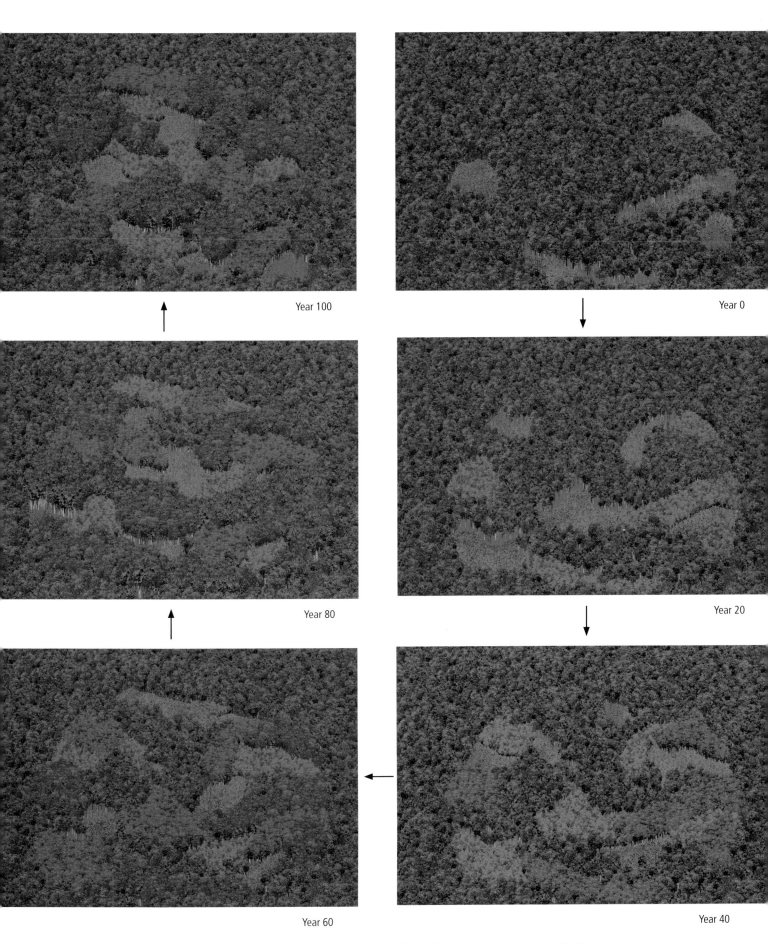

Year 100

Year 0

Year 80

Year 20

Year 60

Year 40

Fig. 38. Group selection in a 20-acre northern hardwood stand over 100 years. Every 20 years, approximately one-fourth of the stand is removed in about half-acre patches with some tree-by-tree removals between groups. Images by Anna M. Lester.

primarily of beech and some sugar maple (fig. 39a and b), which limits both the diversity and revenue potential of the future stand. With groups, there is a significant added component of yellow and paper birches, white ash, pin cherry, and miscellaneous other tree species and shrubs (fig. 40). In addition, there is some efficiency in removing groups of trees.

There is another option that should be mentioned. On large properties, 500 or more acres, it is quite possible to do some harvesting every year. With a 20-year return cycle to the same area, this would mean that group selection could be applied to about 5 percent of the property each year. In 20 years, the whole property would have been covered with a harvest operation, and we would be back where we started (fig. 41). This option provides a more regular

Fig. 39a and 39b. Single-tree selection in northern hardwoods often produces abundant beech regeneration in the understory and an abundant nut (hard mast) crop, which is an important wildlife food, especially for black bear. Photographs: (a) USDA Forest Service, (b) Mariko Yamasaki.

income and it provides a more consistent array of wildlife habitat conditions.

How much income would be generated using group selection? Experience shows that northern New England stands produce up to one-half cord per acre per year on a regular basis. Let's be conservative and call it one-third cord. On a thousand-acre property, this means an average harvest of 300 or more cords per year including all products—veneer, sawlogs, pulpwood, fuelwood, etc. In a well-stocked stand with a good proportion of larger trees, it is possible that 20 to 40 percent of the harvest would be sawlogs/veneer—and the proportion should get higher over time. Probably the most important wildlife reason to use this approach is to provide some early-successional and young forest habitat continuously on the property, although in small patches. Loss of such habitat is the most pressing wildlife habitat concern in New England, and using group selection every 5 to 10 years would be somewhat beneficial to a range of species such as towhees and snowshoe hares, among many others.

CHANGES IN STAND CONDITIONS. With the uneven-aged management scenario, the coniferous stand would maintain a high proportion of

Fig. 40. The greater diversity of species (including yellow birch, paper birch, white ash, raspberry, and pin cherry) produced by group cuts has an overall greater timber and wildlife potential than single-tree selection. This group cut is about 10 years old. Photographs by Mariko Yamasaki.

coniferous species coupled with small patches of young conifers and deciduous tree species and shrubs. In this sense, the stand composition would be very similar to that found under the no-harvest scenario where the small patches occurred at very irregular time intervals through natural disturbance. A component of cavity trees and downed woody debris would be maintained by carefully leaving some low-quality and risky trees during the harvest operations, but the proportion would be lower than under the unmanaged scenario. Understory trees and shrubs would be somewhat higher with uneven-

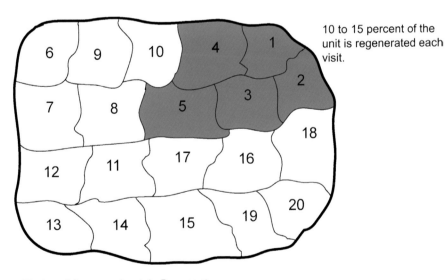

10 to 15 percent of the
unit is regenerated each
visit.

Each unit is approximately 5 percent
of the tract. Factors such as access,
overmaturity, and marketability can drive
the layout pattern.

Fig. 41. An uneven-aged group selection schedule to create regeneration patches across a large ownership. In this example, the ownership is divided into 20 units. Every year, 10 to 15 percent of each unit is cut. This schedule ensures that early successional patches are always present on the property. Clumping units concentrates the area and effectiveness of early successional habitat.

aged management than with no harvest, due to the extra light penetrating through the canopy when trees were harvested.

The northern hardwood stand would develop quite differently with group selection than with no harvest. Recall that under the no management scenario, most of the young trees would be beech and some sugar maple. The groups (10 to 15 percent of the area at each entry) would become dominated by a wide range of trees and shrubs including birches, ash, pin cherry, some aspen, and raspberry. However, the remainder of the stand would have a closed canopy but with groups of smaller trees (saplings and poletimber) that were developing in earlier group cuts.

Cavity trees and dead woody debris would be maintained by leaving some trees with these potentials. The reserve area, which would remain unharvested, would develop in the same way as described under the no-harvest scenario.

From a distance, the landscape would look quite homogeneous over time. The small openings produced by group selection would add a rough texture to the landscape (fig. 42), and from certain high-elevation vantage points, new group-selection cuts might be obvious for a few years.

There are many species that reside in unmanaged and uneven-aged forests, and others with large home ranges that use such stands. The distributions of many amphibians such as red efts and redback salamanders, small mammals, and resident birds are similar in these stands (fig. 43). There are other species, however, that occur primarily or only in specific stand conditions. The following gives examples of species that are characteristic, regularly occurring, or that are attracted to specific conditions such as group cuts, cavity trees, or soft and hard mast crops, among others.

WILDLIFE HABITAT CONDITIONS. Under an uneven-aged option, canopy gap size would be small (a half-acre in hardwoods; a tenth of an acre in softwoods) and occur on less than 15 percent of the treated stand or landscape during any 20-year period. The wildlife habitat value of these gaps is

Fig. 42. From a distance, group selection cuts provide a rough texture to the forested landscape, which becomes less distinct a few years after the cutting. From a high vantage point, new group cuts may look somewhat obvious for a few years. Photograph by Christine A. Costello.

Fig. 43. **Below,** A red eft, the terrestrial juvenile form of the red-spotted newt, travels widely in the forest before returning to a pond and resuming its aquatic life. The red eft and several other species use coarse woody debris for shelter. This large white pine and several others fell on Mount Toby, Massachusetts, in the 1955 hurricane. Photograph by Richard M. DeGraaf.

short-lived and would last for about 8 to 10 years after the cut. Therefore, landowners might consider distributing this temporary habitat throughout the 20-year span of the management interval, as compared with completing all treatments at one time. Low-intensity road and skid trail access would be needed to periodically access these stands. There would be few if any changes to the overall forest habitat quality (i.e., no forest fragmentation and minimal invasive species exposure) with a system of narrow roads/trails accessing these stands and they would enhance recreational use of the property (fig. 44).

Table 2 shows that a partial to closed overstory canopy would exist throughout the landscape under this management scenario. Structural habitat features such as exposed perches, small patches of open canopy, some soft mast, and an ephemeral herbaceous layer would be apparent in most freshly cut groups for a few years. The rest of the stand would have an essentially closed canopy and midstory that would not permit sufficient sunlight to reach the ground and stimulate those vegetative responses. Hardwood inclusions in softwood stands and softwood inclusions (which add species—primarily birds—that otherwise would not be present) in hardwood stands could be maintained with group selection. Large cavity trees and coarse woody debris can be developed in stands by retaining a sufficient number of trees per acre in small retention patches and carefully marking between groups. Densities of large cavity trees and coarse woody debris would probably be less than those in the no-management scenario, but the overall wildlife communities, except for the early-successional species associated with the group cuts, would be similar to those in unmanaged stands.

Treating the lower elevation softwood sites using uneven-aged group selection, one would expect minimal herbaceous groundcover and soft mast response in the treated areas, but a strong woody regeneration response of mid-tolerant to tolerant softwood and hardwood species. Such patches of early successional, mostly hardwood, habitat attract such species as woodcock, cuckoos, alder flycatchers, veery, chestnut-sided warbler, rose-breasted grosbeak, meadow vole, and snowshoe hare (fig. 45), among others. These species will normally persist until the patch grows out of the brushy stage and then occur in new patches as they are cut. Soft mast (e.g., serviceberry, strawberry, raspberry, pin cherry, and blackberry), and a substantial herbaceous groundcover would rarely occur. Large cavity trees could be present with careful marking but at lower density than in no-management scenarios. The amount of coarse woody debris that would be present depends upon how much marking between groups occurred. Marking just the mature trees next to skid roads and trails would retain a higher amount of coarse woody debris than marking the entire area between groups. The wildlife community associated with uneven-aged coniferous stands will normally be more diverse than that in the unmanaged stand because of the additional species associated with the group cuts.

Managing northern hardwood sites using uneven-aged group selection would produce some herbaceous groundcover and soft mast response in the treated areas, as well as a strong woody regeneration response of mid-tolerant to tolerant hardwoods, and even a little aspen. Uneven-aged northern hard-

Fig. 44. Typical low-intensity trail in northern New England. Photograph by Mariko Yamasaki.

Fig. 45. Snowshoe hare pelage turns white in winter; the hare is an important prey species for the northern goshawk, fisher, bobcat, and lynx. Photograph by Mariko Yamasaki.

Fig. 46. Chestnut-sided warblers are among the first birds to breed in hardwood clearcuts and group cuts. They abandon the site after about 10 years, when dense foliage is no longer present within 3 feet of the ground. Photograph by David I. King.

wood stands harvested with group selection (5 to 10 year entry) have fairly diverse wildlife populations compared to unmanaged ones. As in the lower elevation softwood stand, the patches of early successional growth — shrubs and seedlings — provide habitat for species that would otherwise not occur. Some typical ones include chestnut-sided warbler (fig. 46), American redstart, mourning warbler, common yellowthroat, and snowshoe hare. More migratory than resident species nest in these patches. Deer and moose would browse these young patches. If the group cuts are very small — one-tenth to one-fifth of an acre or so — the patches of regeneration will be very small and indistinct, and early successional birds such as mourning warbler, indigo bunting, and eastern bluebird will not occur in them. Such small cuts would not increase the intolerant tree and shrub composition. In untreated parts of the stand between patch cuts there would be no herbaceous groundcover and soft-mast-producing plants. There would be a visible midstory, a partial to closed canopy, and a limited amount of beech regeneration over time. Hard mast would essentially be all beechnuts; a small amount of oak might be present producing some acorns on south-facing slopes and ledge, and if red spruce is present, periodic cone crops would provide seed. The density of large cavity trees would be reduced in the management process but could be maintained with careful retention guidelines. How much coarse woody debris would be present in the subsequent stand would depend upon how much marking between groups occurred. As in the softwood sites, marking just the mature trees next to skid trails would retain a higher amount of coarse woody debris than would marking the entire area between groups.

Since there is no management on the higher elevation softwood sites, we would expect the same types of habitat described earlier in the "no management" option.

Even-Aged Management Scenario

This is the most disturbance-intensive scenario, best suited to the landowner who wants a high level of efficient timber harvest coupled with a desire to provide habitats for a diversity of wildlife species, especially those requiring large patches of early-successional habitat and young growth. The scenario in the northern New England situation consists of a range of clear-cut sizes from 5 to 40 acres. We'll use 20-acre clear-cuts in the northern hardwood stands and 20-acre shelterwood cuts in the lower elevation conifers as examples. A shelterwood cut is used in coniferous stands because a clear-cut here would largely convert the stand to hardwoods. No harvesting occurs in the reserve areas. In both stands, we assume that it will take about 100 years to grow a mature tree (the so-called rotation age). So, every 20 years, one-fifth of the northern hardwood area would be clear-cut and one-fifth of the coniferous area would be shelterwood cut. Both systems produce new stands of trees that are about the same age — hence the term even-aged management.

The clear-cut is easily described — everything is removed except for perhaps islands of residual trees that have some unusual habitat feature (large

Fig. 47. A typical clear-cut with an inclusion in which all trees were removed in northern New England. Commonly, patches of reserve trees are left in the clearcut and along watercourses. Photograph by Mariko Yamasaki.

cavity trees, for example) (fig. 47). A traditional shelterwood cut, as the name implies, leaves a component of overstory trees—perhaps one-half to two-thirds of the stand—to shade the young regenerating trees (fig. 48). Then, about 10 years later, most of these overstory trees are removed, leaving the young stand of seedlings and saplings to develop. Under both the clear-cutting and shelterwood options, when the young stands of trees reach merchantable size—in perhaps 50 to 60 years—they commonly are thinned to concentrate the growth on the best trees. Then at age 100, the trees would be mature enough for another clear-cut or shelterwood cut.

Fig. 48. Example of a shelterwood cut in hemlock-spruce about 5 years after the harvest. Note the developing softwood regeneration. Photograph by Timothy L. Stone.

One other option deserves mention: Instead of a complete clear-cutting in northern hardwoods, a low-density "deferred" shelterwood (sometimes called a delayed shelterwood or a shelterwood with reserves) can be used. This option involves harvesting a mature stand so as to leave a sparse, scattered overstory of trees that would be harvested after several decades or left to eventually die (fig. 49). This system regenerates commercial tree species composition similar to clear-cutting; however, there are lesser amounts of pin cherry, aspen, and fruiting shrubs so the wildlife value is usually a bit lower. It is sometimes used in areas where clear-cutting is not aesthetically desirable.

As with the uneven-aged scenario, it is quite possible on large properties to do some cutting every year or few years. For example, one could clear-cut or shelterwood harvest every 10 years on 10 percent of the acreage.

Fig. 49. A deferred shelterwood in northern hardwoods. The residual trees are not abundant, so enough sunlight is provided for the regeneration of a mixture of shade-intolerant (sun-loving) and shade-tolerant species. The residual trees are left in place for a few to several decades. Photograph by William B. Leak.

Timber production under the even-aged scenario would be fairly similar to that under uneven-aged management—perhaps ⅓ to ½ cord per acre per year. However, timber harvesting is commonly more efficient under even-aged management, there is a smaller component of less valuable species (beech and red maple), and the quality of trees may be better under the even-aged option.

CHANGES IN STAND CONDITIONS. The unique feature of the even-aged system is the maintenance of large-scale horizontal diversity—simply meaning that there are large patches (20 acres in this case) of openings, sapling stands, poletimber stands, sawtimber stands, and old forest spread across the landscape. Over time (fig. 50), the entire property becomes a mosaic of early-successional habitat, young and mature even-aged hardwood (fig. 51) and softwood stands (fig. 52), and old forest. Where such a system is used, the greatest forest wildlife habitat diversity occurs.

In the coniferous area of the property, following the initial shelterwood cut (called the seed cut), a young stand of mostly coniferous species (hemlock, spruce, and some pine) should develop as an understory beneath the

Fig. 50. Time sequence of a northern New England forest tract from year 0 to 100 using even-aged management. After 100 years, various seral stages are present across the landscape; however, the area in which they occur shifts over time as stands regenerate and mature while others are harvested. Most upland wildlife species in northern New England would find suitable habitat in this landscape. Images by Anna M. Lester.

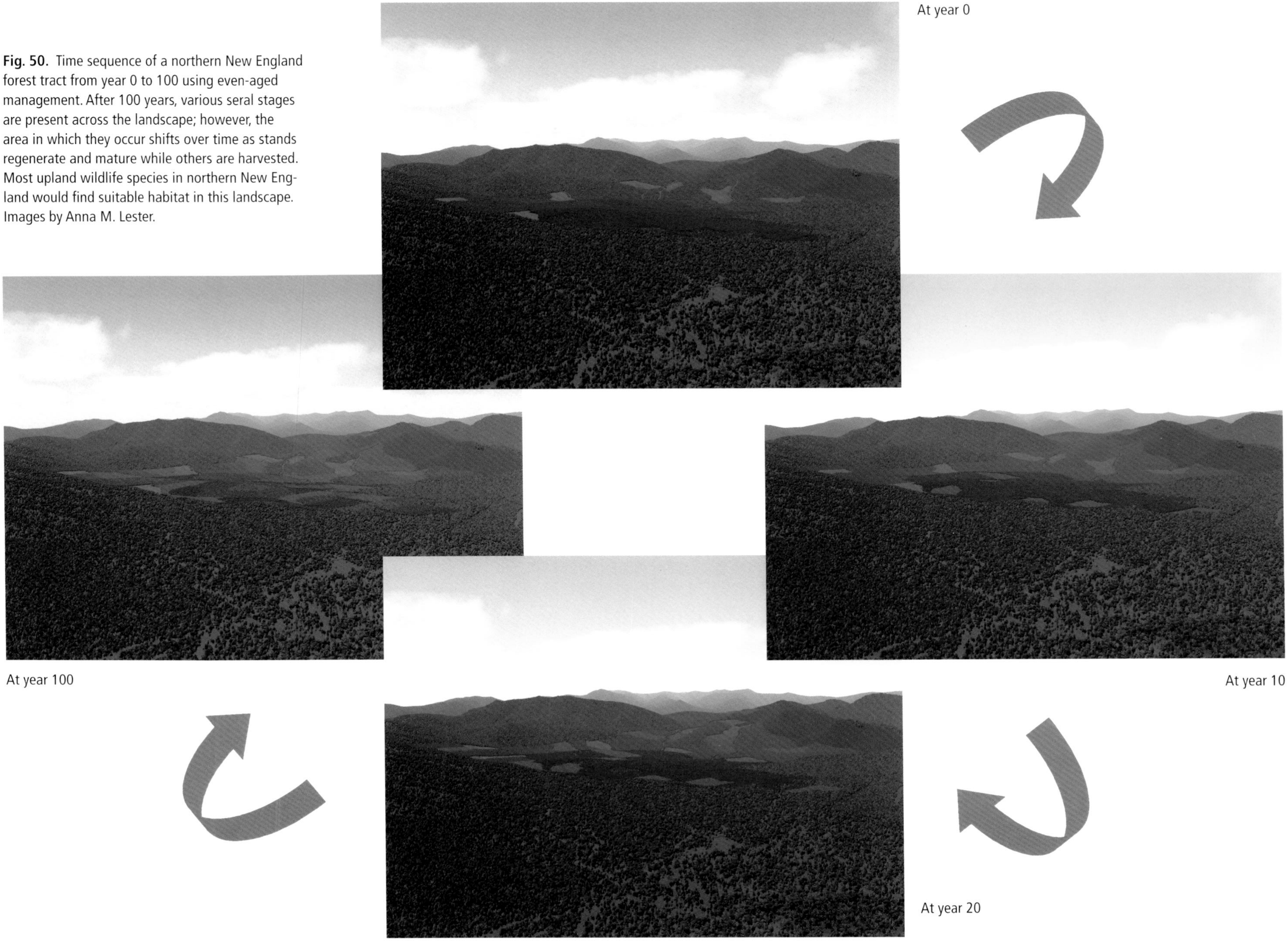

At year 0

At year 10

At year 20

At year 100

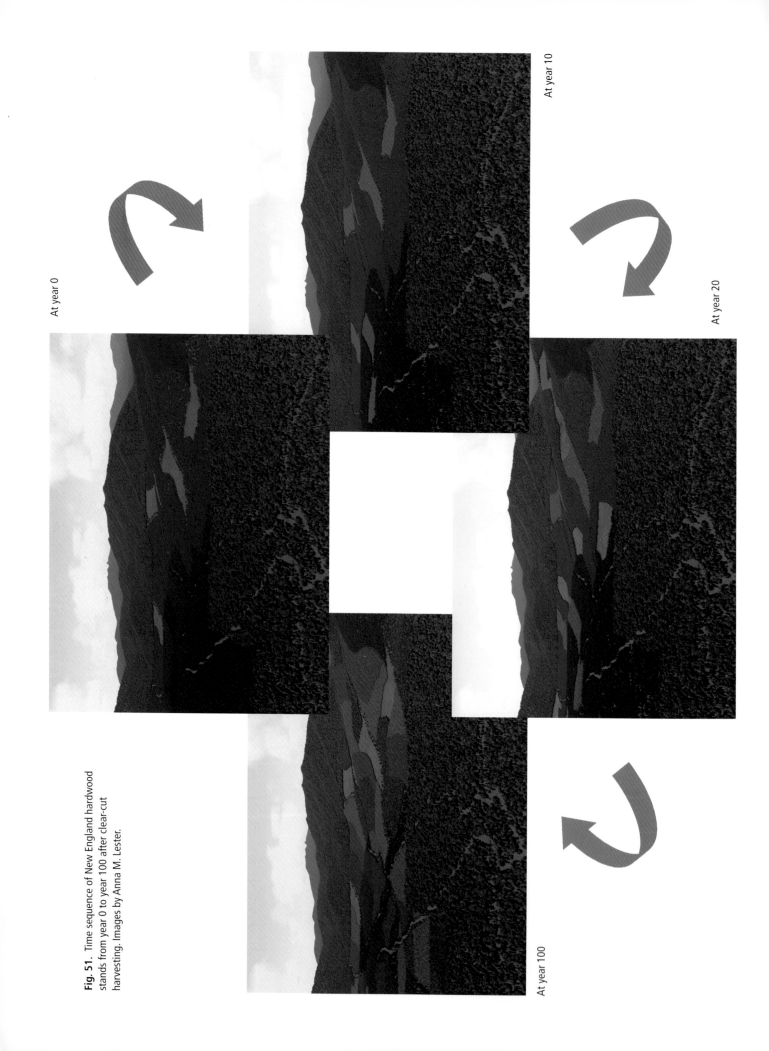

Fig. 51. Time sequence of New England hardwood stands from year 0 to year 100 after clear-cut harvesting. Images by Anna M. Lester.

At year 0

At year 10

At year 20

At year 100

Fig. 52. Time sequence of northern New England softwood stands from year 0 to year 100 using a shelterwood system. Images by Anna M. Lester.

At year 0

At year 10

At year 20

At year 100

sheltering overstory. Note that these initial shelterwood cuts appear as areas of scattered large trees (fig. 53). Then, 10 years later comes the so-called overstory removal, which releases the understory to full sunlight. This removal entails a significant amount of harvesting disturbance, although equipment is now available to prevent too much destruction of the existing understory. In the disturbed areas, a mixture of birches, pin cherry, aspen, and shrubs regenerates, providing for a deciduous component to the developing young stand. Over time, the deciduous component becomes somewhat reduced due to mortality of the short-lived shrubs and pin cherry and the harvest (thinning) of paper birch and aspen at 50 to 60 years of age. At maturity (100 years), the stand has a high percentage of coniferous species, including a component of cavity trees, dead snags, and downed woody debris provided for by reserving trees with these potentials during the thinning operations.

The clear-cut northern hardwood stands immediately regenerate to a mixture of birches, ash, pin cherry, aspen, and shrubs—still with a component of the shade-tolerant species (beech and sugar maple) (fig. 54). The raspberries disappear within a few years, the pin cherry dies by about age 30 to 35, the aspen and paper birch would be thinned at age 50 to 60, and the older stands would be comprised largely of yellow birch, ash, sugar maple, and beech. Since these stands tend to develop a closed canopy at an early age, understory development is normally quite sparse. However, fairly intensive thinning at ages 50 to 60, and perhaps about age 80 will help create or maintain a fairly dense understory and also provide some timber revenue. As in the coniferous stand, careful timber marking practices during the thinning operations are required to maintain a component of cavity trees, dead snags, and woody debris.

WILDLIFE HABITAT CONDITIONS. This "even-aged" option creates large openings in both hardwoods and softwoods. Such openings would occur on less than 15 percent of the treated stand or landscape during any 20-year period. This option gives landowners the best chance of regenerating intolerant as well as mid-tolerant and tolerant species and creates the widest range of wildlife habitat conditions. We could expect a substantial herbaceous groundcover and soft mast response, given the larger size of the created gaps. The primary value of these gaps for wildlife would last for 8 to 10 years after the cut (fig. 55). The greatest wildlife benefit is reached by distributing this ephemeral habitat throughout the 20-year span of the management interval, rather than completing all treatments at one time. Note in table 2 that, under even-aged management options, the overstory canopy can be very open to partly open in recently harvested stands. Where a partial canopy is left, the crowns of these remaining trees continue to develop as the new stands grow. This process can affect the presence or absence of structural wildlife habitat features and, in turn, the wildlife species that will utilize the stand. The more open the residual canopy, the more exposed perches, herbaceous growth, shrub habitat, and fruit production will be produced.

High and low exposed perches are used by forest raptors for hunting and by birds such as the chipping sparrow or olive-sided flycatcher for song

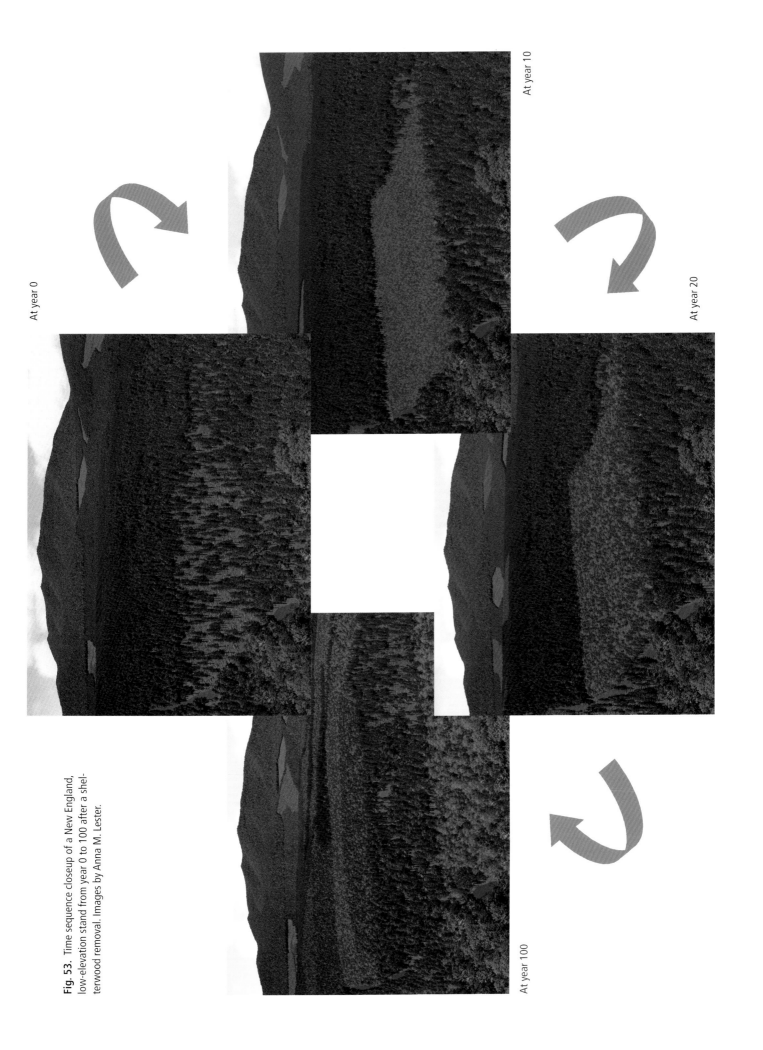

Fig. 53. Time sequence closeup of a New England, low-elevation stand from year 0 to 100 after a shelterwood removal. Images by Anna M. Lester.

At year 0

At year 10

At year 20

At year 100

Fig. 54. Time sequence closeup of a New England mid-elevation hardwood stand from year 0 to year 100 after a clear-cut. Images by Anna M. Lester.

At year 0

At year 10

At year 20

At year 100

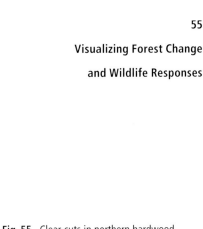

Fig. 55. Clear-cuts in northern hardwood stands regenerate quickly and successfully to a mixture of deciduous species, plus a component of shrubs such as raspberry. This clear-cut is about 20 years old and its early-successional value for most species is waning. It is, however, prime ruffed grouse habitat. Photograph by Mariko Yamasaki.

perches or for sallying to feed on flying insects. Such perches are available in clear-cuts and shelterwoods until the forest regeneration grows tall enough to preclude their use as hunting or song perches. Maintaining selected overstory inclusions and large cavity trees in leave or retention patches or by careful marking helps to build structure in the new stand. Hard mast availability in newly harvested clear-cuts will be low until the new stand is old enough to start producing seed again. Soft mast usually is plentiful in newly regenerating stands as long as there is no overstory shading. Herbaceous forage and soft mast like strawberries disappear in clear-cuts when the dense woody regeneration of the shrub layer shades the strawberries out, normally in 2 to 5 years. Raspberries and blackberries stop producing soft mast in 5 to 10 years when the woody regeneration grows tall enough to shade them out. Pin cherry is short-lived; it grows into the overstory canopy and produces soft mast for a wide range of species including foxes, black bear, and cedar waxwings, among many others, from about age 15 to 30 or 35, when it dies. Midstories develop in time but are not usually present in recently harvested stands unless lots of small hardwood and softwood "whips"—large saplings and small poles— remain standing. Leaving too many of these stems will influence the species composition of the regenerating stand and will reduce the development of early successional herbs, shrubs, and fruit, or shorten the time they are available. Such whips should normally be removed to favor intolerant and mid-tolerant tree species (tolerant species will still comprise half of the regenerating stand). A variety of tree and shrub species ensures that a variety of plant and insect food, nest sites, and cover are present.

In softwoods, traditional shelterwood treatments create a brief herbaceous groundcover and soft mast response in treated areas, as well as a strong regeneration response of intolerant, mid-tolerant, and tolerant softwood and hardwood tree species. In addition to seeds from coniferous cones, any beech present would periodically produce a hard mast crop every 3 to 5 years; few red oak acorns would be produced on softwood sites. We would expect low hard mast production over time, and that primarily beech. Large cavity trees would be present in both hardwood and softwood stands with careful retention marking, and retention of large coarse woody debris would be present with careful marking.

A distribution of even-aged hardwood stands provides habitat for the most diverse wildlife populations because optimal habitat conditions for many species occur predictably on the property in a shifting mosaic. There are four groups of breeding birds, for example, inhabiting seedling, sapling, poletimber, and mature stands. The general pattern is high diversity in regenerating stands, lowest in poletimber stands, and high again in mature stands. In New England, no species need stands older than silvicultural rotation age. The same species occur in even-aged sawtimber stands as in stands twice as old. Characteristic bird species in each stand are: ruby-throated hummingbird, common yellowthroat, chestnut-sided warbler, cedar waxwing, and willow flycatcher in regenerating seedling stands; redstart, veery, blue jay, and ruffed grouse in sapling stands; red-eyed vireo, wood thrush, ovenbird, and scarlet tanager in poletimber stands; and least flycatcher, black-throated blue warbler, and hairy woodpecker in mature stands. Some species occur at the edges of clearcuts. Baltimore oriole and olive-sided flycatcher are examples. Others need a range of stand ages to nest and feed — goshawks nest in mature trees but feed on grouse, hare, and other species associated with younger stands. Ruffed grouse generally reside in sapling stands but need herbaceous openings for broods to feed on insects and mature stands for winter feeding on buds, especially aspen.

In untreated parts of hardwood sites, we would expect a lack of herbaceous groundcover, soft mast elements, and a closed canopy. Hard mast diversity would consist primarily of beech nuts over time with the decline in acorn production. Large cavity trees would be present in both hardwood and softwood stands with careful retention marking. Retention of large coarse woody debris would be present with careful marking but probably would not accumulate largely due to the rapid decomposition of wood in the northeastern United States.

In both hardwoods and softwoods, shelterwood cutting has many of the habitat advantages of clear-cutting but commonly the species that occur only in very open regenerating stands, such as bluebird and olive-sided flycatcher, and species associated with closed-canopy mature stands — wood thrush, Swainson's thrush, and great crested flycatcher, for example, may not be present, though most young forest species are usually present. Depending on how much cull was left in the overstory, woodpeckers and other cavity nesters may be common or uncommon.

Although mammals are generally not as habitat specific as birds, regenerating clear-cuts provide abundant summer fruit for foxes, raccoons, and black bears and winter browse for deer and moose. Regenerating and sapling stands provide secure year-round cover for snowshoe hares, as well as for species in decline throughout the region due to lack of such habitat. Mature stands provide beech and oak mast for deer, bears, squirrels, chipmunks, mice, and other species. Clearcutting cannot be practiced everywhere, but where possible, it is an especially valuable wildlife habitat management practice.

Summary

In extensively forested northern New England, there are several basic differences in stand structure and, in turn, wildlife communities that result from uneven-aged and even-aged management. Use of uneven-aged management with group cutting provides habitat components such as patches of early successional habitat and young forest predictably that generally would not be present, or would occur rarely, in the absence of management. Wildlife communities, therefore, are somewhat more diverse in forests managed with group cutting than in those that are left unmanaged.

Even-aged management with clear-cut regeneration provides large patches of early successional habitat, young forest, and mature and old forest conditions in a shifting mosaic over time. Such management provides habitat for the most diverse wildlife community, and maintains forest and wildlife diversity through time. Most of the wildlife diversity is associated with seedling and sapling stands; once beyond the poletimber stage, stands have about the same wildlife species whether they are even-aged sawtimber or old forest.

Southern New England

Conditions on this tract in southern New England, about 2,100 acres in size, were adapted from cruise data provided by the Massachusetts Division of Fisheries and Wildlife. There are two forest types represented (fig. 56), in stands about 7 to 8 acres in size somewhat intermixed on the landscape: oak–pine (primarily red maple, red and white oaks, beech, black and white birch, with a softwood component of up to 25 percent in white pine and hemlock; fig. 57), and softwoods (of which up to 75 percent is white pine and hemlock) (fig. 58). The oak–pine stands often occur on soils that are fairly deep and moist, although most of these species also tolerate dry conditions. The softwoods often occur on soils that are drier than usual (sand/gravel deposits or shallow

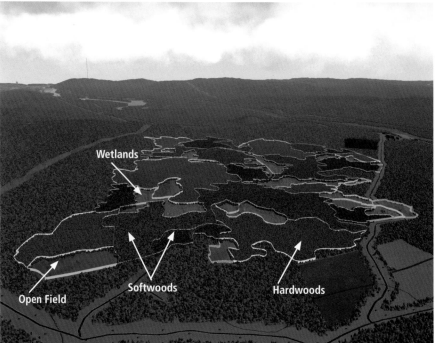

Fig. 56. A typical southern New England landscape with two forest types represented: hardwoods dominated (oak–pine) and softwood dominated (white pine/hemlock). There are also nonforest features present: wetlands and open fields. Images by Anna M. Lester.

bedrock) or wetter than usual (shallow, poorly drained hardpan). All of the initial stands are approaching maturity, supporting trees up to 18 to 20 inches in diameter and occasionally larger. In addition, there are some intermixed wetlands and field (fig. 59a and b), and the tract is bordered by some open fields and residential properties.

In many parts of southern New England, invasive exotic plants such as glossy buckthorn, Japanese barberry, multiflora rose, and oriental bittersweet are common, especially in fields and woodlots near residential areas. They are

Fig. 57. Left, Typical oak–pine stand in southern New England containing red oak, red maple, black birch, white pine, and other species. Typically, white pine stands head and shoulders above the other species. Photograph by Mariko Yamasaki.

Fig. 58. Below. Typical, old hemlock–white pine stand in southern New England. Hemlock is one of the longest-lived species in New England, reaching an age of 400 to 500 years old. Photograph by Richard M. DeGraaf.

Fig. 59a and 59b. Typical wetlands and old fields in southern New England forests. Photographs: (a) Richard M. DeGraaf, (b) John Scanlon.

well established and difficult to control. Some, including Japanese barberry, multiflora rose, autumn olive, and tatarian honeysuckle, have high wildlife food and/or cover values. Their presence need not be viewed as an obstacle to forest management because they only occasionally achieve densities that affect tree regeneration, and cannot be kept out by avoiding forest management. This in no way implies that invasive plants should be planted, maintained, or encouraged—they should not, and should be removed where possible. In most forest management situations, invasives can be treated the same as native species—hay-scented fern and beech—that in some situations may become dense enough to prevent regeneration. The proportions of both invasives and native species can be controlled silviculturally to regenerate desired species. Let's keep invasive species in perspective: We want to maintain productive, working forests on the landscape as wildlife habitat. Treating them as somehow degraded because of invasive species will only lead to commercial or residential development and then all habitat values are lost.

In the following, we discuss three forest management scenarios: no management, uneven-aged management, and even-aged management.

No Management Scenario

In this scenario, the stands are left alone without any harvesting activity; it is best suited to the landowner who wants to maintain the property as it is and values the property primarily for recreation, natural beauty, and privacy, or as a screen from adjacent roads and abutters. In areas where there are also open fields and wetlands, the relatively closed-canopy forest offered in this scenario will provide wildlife habitat conditions that are similar to those common in heavily forested landscapes in northern New England. While landowners want to minimize forest disturbances, we realize that most old fields, pastures, or agricultural land will probably not be maintained over time, and some will be converted to a residential or commercial land use. Under this no-management scenario, the forest from a distance would continue to look about the same over time (see fig. 56); however, there would be some natural changes as described later.

CHANGES IN STAND CONDITIONS. The oak–pine stands would first experience an increase in the dominance of overstory red oak, which is a fast-growing, crown-aggressive species, and overstory pine (fig. 60a and b). Even red maple, an aggressive species in its own right, generally does not grow as fast as well-established red oak. However, with no disturbance over time, we would not expect the oaks to regenerate very successfully, and the understory would become dominated by red maple, beech, some hemlock and pine, and shrubs such as beaked hazel. As the oaks mature and die, the understory tree species would dominate. Damage to the overstory oaks, from the gypsy moth, for example, would hasten the trend. We would not expect, however, large open areas of dead or windthrown trees, or naturally occurring areas of young seedling or sapling stands. Standing dead snags, cavity trees, and dead material on the ground would be abundant.

Most oak–pine stands in southern New England occur on lands with a long history of agriculture, which has influenced the soils and species composition. As a result, changes in species composition (i.e., natural succession) is variable—and somewhat difficult to predict.

The softwood stands would tend to develop a patchy understory with more hemlock and less pine. We use the term patchy because undisturbed softwood stands cast so much shade that little understory develops unless there is some disturbance to the overstory. However, we would anticipate that natural disturbance through windthrow and ice damage would be more common here than in the hardwood-dominated forest types. In addition, there are growing concerns about hemlock mortality from an exotic insect, the hemlock wooly adelgid. Large, open patches in this forest type of up to several acres in size tend to regenerate to birches, especially black birch, other sun-loving tree species, and shrubs. Dead snags and downed material would be abundant, more so here than in the deciduous and mixed-wood types because the conifers are prone to windthrow disturbance and their decay rate is much slower.

Fig. 60a and 60b. Red oak and white pine often become the dominant trees in oak–pine stands. However, regeneration of both species requires care and attention to seed years and harvesting methods. Photograph by William B. Leak.

WILDLIFE HABITAT CONDITIONS. Habitat conditions in this landscape will change dramatically as fields and shrublands revert to forest. Habitat conditions for early successional species will be replaced by others that are characteristic of more extensive mature forest. As shown in table 3,

Table 3
Management Scenarios in Southern New England

Within-stand feature	Example wildlife species	No management	Uneven-aged management		Even-aged management		
			Single-tree	Group/patch	Thinning	Shelterwood	Clear-cut
Closed canopy	Sharp-shinned hawk, great horned owl, hairy woodpecker, northern flying squirrel, porcupine	Tree-sized gaps	Tree-sized gaps		X	Canopy closes in time	Canopy closes in time
Partial canopy	Mourning dove, black-billed and yellow-billed cuckoos, red-bellied woodpecker, Nashville warbler, black-and-white warbler, Baltimore oriole, bobcat			0.1–0.5 acre gaps	X	When first cut	
Open canopy	Northern flicker, willow flycatcher, eastern bluebird, New England cottontail						When first cut
High exposed perches	Turkey vulture, osprey, red-tailed hawk			X		X	X
Exposed perches	Eastern phoebe, eastern kingbird, eastern bluebird, indigo bunting, chipping sparrow			X		X	X
Hardwood inclusions in softwood	Red-eyed vireo, white-breasted nuthatch, brown creeper, black-throated blue warbler, Baltimore oriole, gray squirrel, white-footed mouse, white-tailed deer	X	X	X	X	X	X
Softwood inclusions in hardwoods	Red-breasted nuthatch, golden-crowned and ruby-crowned kinglets, hermit thrush, northern parula, black-throated green warbler, red squirrel, white-tailed deer	X	X	X	X	X	X
Large cavity trees	Wood duck, barred owl, eastern screech owl, pileated woodpecker, northern long-eared bat, porcupine, raccoon, gray fox, fisher	Abundant	X*	X*	X*	X*	X*
Hard mast	Wood duck, ruffed grouse, wild turkey, blue jay, eastern chipmunk, gray squirrel, northern and southern flying squirrels, deer and white-footed mice, black bear, white-tailed deer	X	X	Not immediate	X	X	Not immediate
Soft mast	Northern cardinal, brown thrasher, cedar waxwing, meadow jumping mouse, red fox, black bear, raccoon			X		X	X

Within-stand feature	Example wildlife species	No management	Uneven-aged management		Even-aged management		
			Single-tree	Group/patch	Thinning	Shelterwood	Clear-cut
Midstory	Veery, wood thrush, American robin, solitary vireo, blue-gray gnatcatcher, American redstart, rose-breasted grosbeak	X	X	Not immediate	X	Not immediate	Not immediate
Shrub layer	Black- and yellow-billed cuckoos, whip-poor-will, willow flycatcher, gray catbird, cedar waxwing, yellow warbler, chestnut-sided warbler, rose-breasted grosbeak, eastern towhee, dark-eyed junco, New England cottontail, meadow jumping mouse, black bear, bobcat, white-tailed deer, moose			X		X	X
Herb layer	Eastern smooth green snake, garter snake, ruffed grouse, eastern bluebird, golden-winged warbler, indigo bunting, meadow vole, meadow and woodland jumping mice, white-tailed deer			X		X	X
Coarse woody debris	Spotted salamander, red-spotted newt, redback salamander, northern black racer, eastern milk snake, ring-necked snake, masked shrew, eastern chipmunk, long-tailed weasel, black bear, fisher, bobcat	Abundant	Minimal**	X**	Minimal**	X**	X**

Note: Within-stand features can be provided through various management scenarios in southern New England. Each scenario provides a suite of within-stand features favorable to different wildlife species. This table highlights which within-stand features can be expected under different management scenarios and prescriptions. Note that the wildlife response in large tracts of unmanaged forest over time would be most similar to some form of single-tree uneven-aged management.

*Any timber management activity can reduce the density of large cavity trees. We mark stands to provide a minimum of three to five large live cavity trees per acre in patches and leave strips in addition to all the other uncut large dead trees in the stand. Large cavity trees near water are important structural habitat features and can be easily maintained in riparian buffers.

**Any timber management activity can reduce the recruitment of coarse woody debris on the ground over time. Leaving some large cull trees, particularly hollow ones, in the woods provides denning and shelter features in managed stands.

Adapted from DeGraaf et al. (1992).

under a "no management" option the overstory canopy across the landscape will eventually be closed, and species such as the sharp-shinned hawk, hairy woodpecker, and porcupine would be expected in such conditions. As agricultural fields, old fields, or small shrubby openings fill in with woody tree species, much of the herbaceous vegetation and soft mast—fruit and berries associated with shrub habitats—will be eliminated over time. Other parcels will be converted to residential or other developed land use. Either case results in further losses of early successional habitat for species such as the eastern smooth green snake, whip-poor-will, eastern towhee, and New England cottontail; these and many others will gradually disappear in the absence of management. Any large openings and wetlands or water bodies with high exposed perches would still provide hunting sites for forest raptors such as the red-tailed hawk, or nesting sites for the great blue heron. Likewise, low exposed perches would serve a similar purpose for the eastern phoebe or indigo bunting to hunt insects in larger grassy openings that continue to exist. A no-management option for the forest would minimize small shrubby openings and roadside edges, and would further reduce the availability of herbaceous and shrubby vegetation and soft mast. Species that use herbaceous vegetation and shrubs, such as the black racer, black-billed and yellow-billed cuckoos, brown thrasher, and meadow jumping mouse, among others, would find little cover in these conditions. The only way to retain some early successional species like these if the forest is not managed would be to maintain any existing fields, either by late summer mowing or early spring burning every few years.

The occasional hardwood tree that would blow over or die in a typical oak–pine stand would create a very small, insignificant canopy gap, allowing only tolerant tree species such as beech to regenerate in it. The wildlife community associated with old oak–pine stands would tend to look more like that in hardwood–pine/hemlock stands over time as the oak component declined. Species typically occurring in these stands include the pileated, hairy, and downy woodpeckers, American robin, hermit thrush, red-breasted nuthatch, and black-throated green warbler. Soft mast — fleshy fruits and berries of serviceberry, strawberry, raspberry, pin cherry, and blackberry, for example — and an herbaceous groundcover would not be present. Hard mast-producing shrubs like hazel-nut would occur in the understory. A predominance of red, black, and white oaks would for a while periodically produce substantial hard mast crops every 3 to 4 years, and the proportion of beech would increase over time. Such a stand would be a heavily used fall and winter feeding site for deer and wild turkey. But as the oaks senesce over the next century or two, we would expect a decreased diversity of hard mast crops over time occurring under a no-management option. Large cavity trees would be present in both hardwood and softwood stands, and cavity users like the wood duck, barred owl, pileated woodpecker, porcupine (fig. 61a), and fisher (fig. 61b) would find nesting and denning habitat here. Large coarse woody debris would be present and abundant for species like redback salamander.

We anticipate larger gaps from more frequent natural disturbances (e.g., windthrow and insects) on softwood sites compared to oak–pine sites. On softwood sites with no recent disturbance event, closed canopy conditions would provide habitat for the relatively few but characteristic species associated with older coniferous stands. These include the barred owl, pileated woodpecker, hermit thrush, red-breasted nuthatch, pine warbler, black-throated green warbler, red squirrel, and deer mouse, among others. The sparse understory would be very patchy and mostly hemlock. The overstory canopy is dense and casts such deep shade that little development occurs in either the shrub layer or herbaceous groundcover. Where some recent disturbance event has occurred and several acres of black birch regeneration now exists, species like the whip-poor-will, eastern towhee, and red-backed vole, among others, will find habitat for a while. Since blowdown events occur more frequently on these softwood sites than the oak–pine sites, softwood coarse woody debris

Fig. 61a and 61b. (a) Porcupines commonly den in large, hollow trees, logs, or dry culverts. (b) Fishers travel widely in search of prey such as snowshoe hare, porcupine, beaver, mice and voles, squirrels (red, gray and flying), raccoons, ungulate carrion, and passerine birds. They also eat a variety of apples, berries, and nuts. Photographs: (a) Mariko Yamasaki, (b) Thomas J. Maier.

would also be abundant, and remain on the site longer, as its decay rate is slower than that of most hardwoods.

Uneven-Aged Management Scenario

Much of southern New England is a mosaic of forests, woodlots, various open habitats, wetlands, agricultural land, and housing developments. There are also areas of extensive forest. In much of southern New England, 20 to 30 percent of the landscape is open due to the presence of pastures, orchards, cropland, active management of old fields, and extensive shrub swamps. Therefore, we have developed an uneven-aged scenario in this southern New England tract that would consist of group selection with some harvesting of individual trees between groups. This approach might well suit landowners who want some timber revenue but only a moderate level of harvesting activity. Every 20 years, about one-fourth of the timber would be removed from the entire property using this silvicultural method. The groups in the oak–pine type would be perhaps one-fourth acre in size, and somewhat smaller in the softwood stands. The groups would cover 10 to 15 percent of the area, and the single-tree removals would account for the remainder of the harvest. The valuable timber species in these forest types are red oak and white pine, and in this sequence as well as in the following even-aged scenario, there is an effort to maintain these two species. At a landscape level, the forest would look quite similar to an unmanaged forest (see fig. 56) except with a less even texture and a somewhat different tree species composition. Over time, uneven-aged management in these types should produce about ½ cord of wood per acre per year (all products) of valuable, readily saleable timber—provided that oak and pine are a significant component. As described earlier in fig. 41, there is the option of more frequent entries on a small proportion of the area—for example, an annual harvest on 5 percent of the property.

Fig. 62. These red oak seedlings (with the dead leaves attached) have been released by a group cut in an oak–pine stand. Where patches of red oak or white pine seedlings occur naturally in the understory, group selection can be used effectively to release them, that is, to provide sunlight and space so they can survive and grow rapidly. Photograph by USDA Forest Service.

CHANGES IN STAND CONDITIONS. In the oak–pine stands, the objective of uneven-aged harvesting would be to maintain—to the greatest extent possible—the oak and pine component. The harvesting would be timed to coincide with a good oak and/or pine seed year, which occurs every 3 to 4 years. The harvesting should be done in the fall, after most acorns have fallen, and the logging disturbance should scarify the site to give the oak the best chance to germinate and reduce the competing understory. Even with these precautions, securing a high proportion of oak and pine seedlings can be chancy. In some areas, there may be understories that already contain clumps of oak or pine seedlings/saplings. These stems, which should be quite well established, can be released from overstory shade by careful application of group openings (fig. 62). When making group openings, it is important to remove midstory trees (often red maple and black birch) along with the saw-timber trees. If left, any midstory trees will impede the establishment of the desired understory of new seedlings and released advance regeneration seedlings and sprouts. However applied, the new groups will also contain high

proportions of sun-loving tree species and shrubs — birches and raspberries, for example. During the application of this system, attention must be paid to leaving a component of dead, dying, and cull trees — especially larger ones — to maintain snags, cavity trees, and coarse woody debris over time.

In the softwood stands, the purpose of group selection harvests would be to maintain a high proportion of conifers with some pine component. On some sites, such as sandy or gravelly soils, the pine component will be fairly easy to maintain. On other soils, such as wet hardpan, it will be more difficult. The approach will be to use smaller groups, about one-tenth acre perhaps, to maintain a high proportion of hemlock, along with the pine, and also to minimize the possibilities of windthrow (fig. 63). Likewise, the marking between groups will be light to moderate to minimize risks of windthrow. As in the other types, some attention must be paid to maintaining dead, dying, and defective trees. The small groups will regenerate a minimal proportion of sun-loving tree and shrub species; however, some natural disturbance will continue to maintain these species in the mix.

Fig. 63. In southern New England softwood stands, we recommend small group openings (~1/10 acre) to regenerate hemlock and pine. Photograph by William B. Leak.

WILDLIFE HABITAT CONDITIONS. Under an "uneven-aged" option, canopy gap size would be small (a quarter- to half-acre in hardwoods; a tenth of an acre in softwoods) and occur on less than 15 percent of the treated stand or landscape during any 20-year period. The food and cover values of these openings to wildlife would occur in the first 8 to 10 years of the cut. Individual landowners might consider distributing this short-lived habitat element throughout the 20-year span of the management interval, as compared with completing all treatments at one time. Low-intensity road and skid trail access would be needed to periodically access these stands.

Table 3 shows that a partial to closed overstory canopy would exist throughout the landscape under this management scenario. Managing oak–pine sites using uneven-aged group selection would produce some exposed perches, an herbaceous groundcover, and a soft mast response in the treated areas, as well as a strong woody regeneration response of mid-tolerant to tolerant hardwoods and softwoods. This option would not actively increase the intolerant tree and shrub composition. Oak–pine stands managed over time using uneven-aged group selection methods can have fairly diverse wildlife populations compared with unmanaged ones. Typical species using larger regenerating group cuts include the Nashville warbler, chestnut-sided warbler, dark-eyed junco, and New England cottontail. More migratory bird species than resident species nest in these patches. Deer browse the woody regeneration in these groups, and high deer populations may degrade the quality of forest regeneration. Where group selection cuts are smaller than $\frac{1}{5}$-acre, regeneration patches will be small and indistinct, and early successional birds such as the indigo bunting and eastern bluebird will not nest in them. In untreated parts of the stand between these cuts/patches, there would be little to no herbaceous groundcover and soft mast-producing plants. There would be a visible midstory of more tolerant tree species, a partial to closed canopy, and some beech regeneration. Without a concerted effort to regenerate the oak species (e.g., harvesting after a large seed crop in the late fall; scarifying seed bed to cover acorns; and eliminating any understory in the group cut), hard mast diversity would dramatically decline and consist primarily of beech nuts over time as the oaks decline and drop out of the stand. Beech composition may increase slightly over time. The density of large cavity trees might be somewhat reduced in the management process but could be maintained with careful retention guidelines. How much coarse woody debris would be present in the subsequent stand would depend upon how much marking between groups occurred. Marking just the mature trees next to skid roads/trails would retain a higher amount of coarse woody debris than marking the entire area between groups. Special efforts would be required to maintain the early successional habitat value of old fields and the grassy habitat component of large hayfields through periodic mowing or field reclamation to maintain or enhance the overall wildlife diversity of the property (fig. 64). Except for the possible addition of early successional species, primarily birds, using large groups, the overall wildlife community associated with the uneven-aged forest would differ little if at all from that in the unmanaged condition. Both would provide valuable fall and early winter feeding sites for deer, bear, and wild turkey (fig. 65).

Treating softwood sites using uneven-aged group selection, we would expect minimal herbaceous groundcover and soft mast responses in the treated areas, but a strong woody regeneration response of mid-tolerant to tolerant softwood and hardwood species in these one-tenth-acre patches. Species in these small patches of regeneration might include red maple and black birch, and on colder sites, beech and sugar maple. These species will normally persist until the patch grows out of the brushy stage and then occur in new patches as they are cut. Some individuals may persist until maturity. Shrubs bearing soft mast and an herbaceous groundcover would rarely if ever occur. Large cavity trees could be present with careful marking but at a lower density compared with the no-management scenario. The amount of coarse woody debris that would be present depends on how much wood is marked between group cuts. Marking mature trees only next to skid trails would retain a higher amount of coarse woody debris in the stand than marking the entire area between groups. Given the high degree of shade that occurs in softwood understories, we would expect a minimal shrub layer and herbaceous groundcover in the untreated portions of the stand. Again, the primary wildlife habitat concern would involve special efforts to maintain the early-successional habitat value of existing old fields or other grassy habitats through periodic mowing or other field reclamation activities.

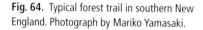

Fig. 64. Typical forest trail in southern New England. Photograph by Mariko Yamasaki.

Fig. 65. Wild turkey feed on hard mast in uneven-aged and mature even-aged stands from late fall to spring. From spring to early fall, they also forage on a variety of plants, fruits, seeds, and insects in open areas. Photograph by William M. Healy.

At year 0

At year 10

At year 20

At year 100

Fig. 66. Time sequence of a southern New England forest tract from year 0 to year 100 using even-aged management. Images by Anna M. Lester.

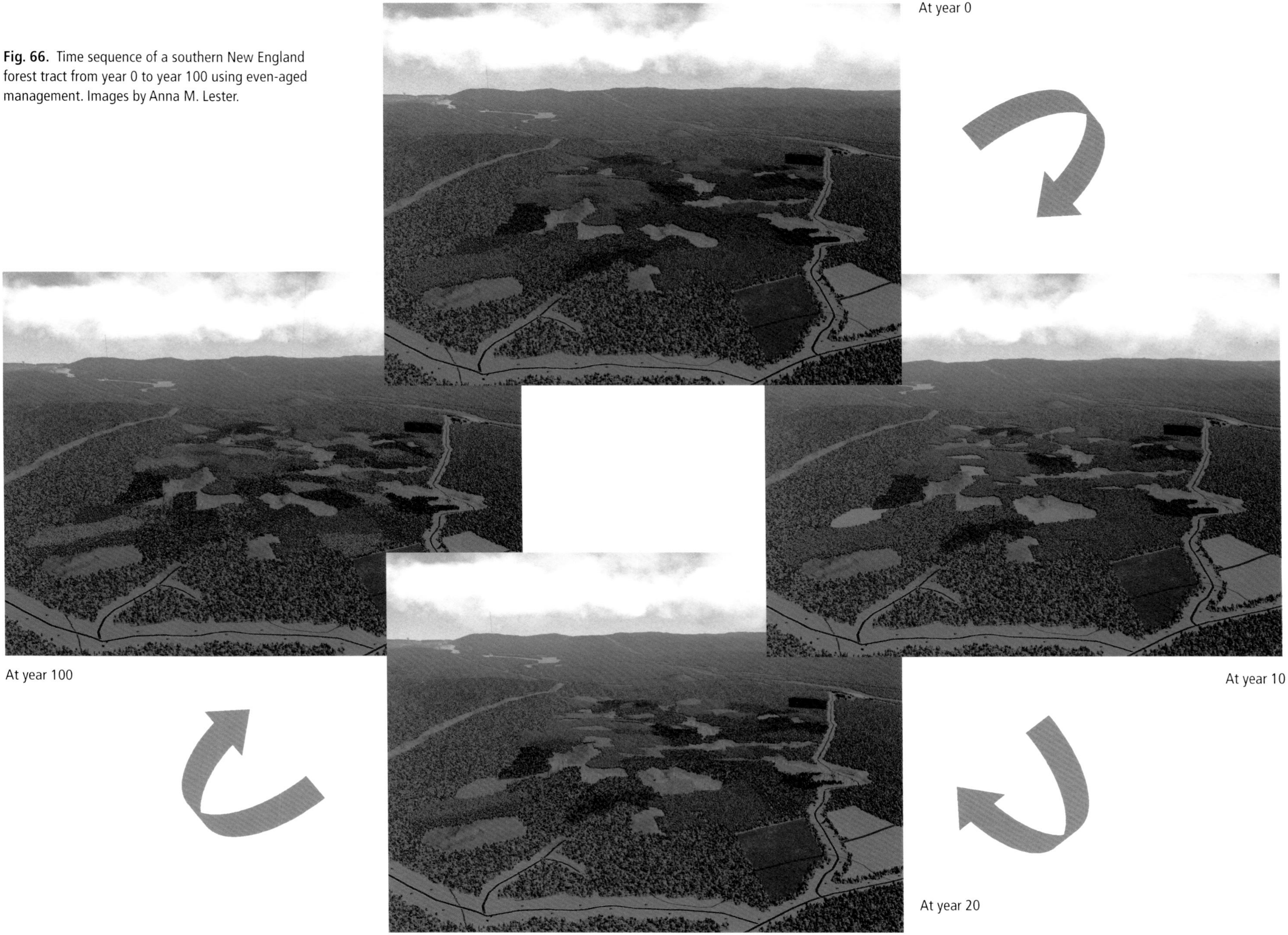

Even-Aged Management Scenario

Where the landscape is a mosaic of forest and open habitats, the need to use even-aged methods to create early-successional wildlife habitats may be somewhat lessened. As in the description for northern New England, however, the even-aged approach provides for the maximum amount of horizontal diversity—areas of young, medium-aged, and older stands in a shifting mosaic across the landscape (fig. 66). And as in northern New England, it provides habitat for the greatest variety of forest wildlife. In the southern New England setting, however, we will use traditional shelterwood cutting in all forest types to meet the needs of the important tree species. Recall that this type of harvest leaves a partial overstory to shade the new seedlings. The harvest areas also are smaller—about 7 to 8 acres, as opposed to 20 acres farther north. Larger or smaller harvest areas could be used. Every 20 years, based on a rotation age of 100 years (the number of years to grow a mature tree), a shelterwood cut would be applied to one-fifth of the area to be managed. Then, after 10 years, the overstory trees would be removed, leaving a young seedling/sapling stand to grow and develop. In a 40-acre tract, a 7- to 8-acre shelterwood cut could be applied every 20 years. At the same time, thinnings could be applied to the younger stands to harvest some wood and improve species composition and quality. In addition, our southern New England even-aged scenario includes a 15-acre patch that will be maintained in the early successional condition by brush-hogging or burning every 10 years (fig. 67).

In the oak–pine stands, the initial shelterwood harvest would leave about half of the overstory canopy, with a high proportion of oak and pine (fig. 68). Most of the smaller trees as well as the undergrowth would be removed by logging disturbance. Timing the initial harvest to coincide with an oak/pine seed year would be useful. In the softwoods, about two-thirds of the overstory would remain to favor the development of a primarily softwood understory of hemlock with some pine; the higher amount of overstory might also limit the amount of windthrown trees—a risk in softwood stands (fig. 69).

CHANGES IN STAND CONDITIONS. In the oak–pine stands, following the initial shelterwood cut, the understory would have an abundance of deciduous trees and shrubs, preferably with a significant oak and pine component, and there would be an overstory of large trees providing for about half canopy closure (fig. 70). After the overstory removal cut 10 years later, the stand would consist of even-aged seedlings and saplings with a broad deciduous mixture, including birches and red maple along with the oak and pine. The oak, although widely scattered at the beginning, should gradually become more dominant as the stand develops.

The intention in the softwood stand is to maintain high proportions of white pine and hemlock. The initial shelterwood cut would initiate the coniferous understory development and also maintain a fairly dense canopy of overstory softwoods (fig. 71). The removal cut after 10 years would have to be done carefully with the right equipment, and with sufficient snow depth, to

Fig. 67. Time sequence of southern New England hardwood stands from year 0 to year 100 with a 10-year entry for a 15-acre hardwood stand. Images by Anna M. Lester.

At year 0

At year 10

At year 20

At year 100

Fig. 68. Time sequence of southern New England hardwood stands from year 0 to year 100 using a shelterwood system. Images by Anna M. Lester.

At year 0

At year 10

At year 20

At year 100

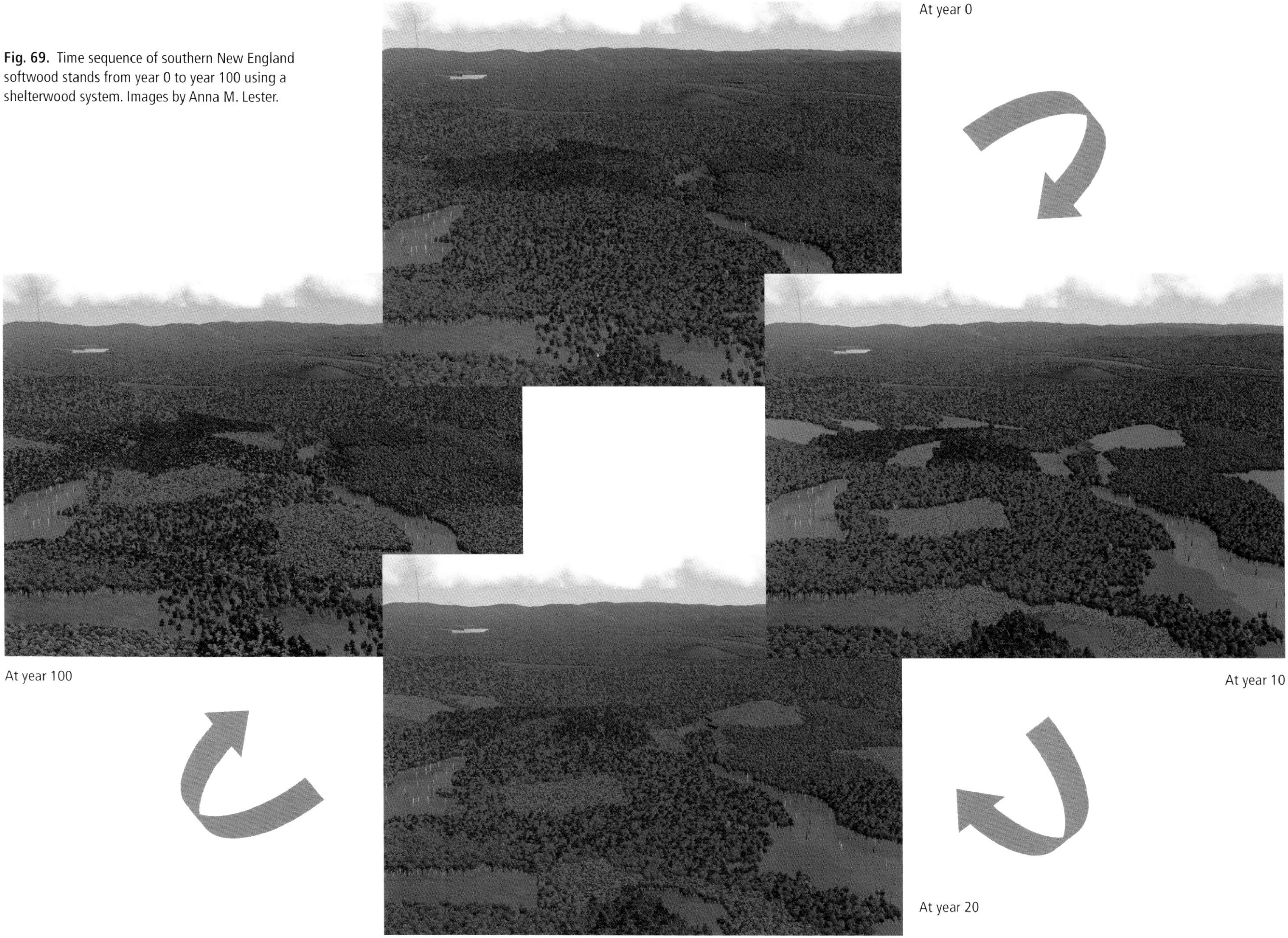

Fig. 69. Time sequence of southern New England softwood stands from year 0 to year 100 using a shelterwood system. Images by Anna M. Lester.

At year 0

At year 10

At year 20

At year 100

Fig. 70. For even–aged management of the oak-pine type, the shelterwood system is the best choice. Note the oak and pine in the understory several years after the initial shelterwood cutting. It is now time to remove the overstory to allow the regeneration to develop and grow rapidly. Photograph by Mariko Yamasaki.

Fig. 71. Shelterwood cut in a white pine–hemlock stand showing the developing conifer understory. Photograph by William B. Leak.

Fig. 72. Short-rotation, early-successional management would periodically regenerate aspen, as well as other shade-intolerant trees, shrubs, herbaceous plants, and grasses. This practice creates dense shrubby cover, soft mast, and an herbaceous cover—all important wildlife habitat features. Photograph by Mariko Yamasaki.

help maintain the softwood component. Although we have assumed a single removal cut, sometimes in softwood stands the overstory removal is done in two operations several years apart.

We can also create short-rotation patches adjacent to wetlands and other nonforest types to improve early-successional habitat. The short-rotation, early-successional stand will begin with a thick cover of birches (including gray birch), aspen, red maple and oak sprouts, and shrubs (fig. 72). If the aspen component is substantial, this species—the fastest grower—will gradually dominate the stand. Otherwise, the stand will continue with very mixed composition. Softwoods will be minimal due to the heavy competition from the deciduous species.

WILDLIFE HABITAT CONDITIONS. This even-aged option creates moderately large openings in both oak–pine and softwoods. Such openings would occur on less than 15 percent of the treated stand or landscape during any 20-year period. This option gives landowners the best chance of regenerating intolerant species such as aspen and birch, as well as mid-tolerant and tolerant species such as oaks, white pine, and hemlock, and creates the widest range of wildlife habitat conditions. Normally, a substantial herbaceous groundcover develops and soft mast-bearing shrubs and small trees proliferate, given the larger size of the created openings. The primary value of these early-successional patches for wildlife would last for 10 to 12 years after the cut. The greatest wildlife benefit is reached by distributing these short-lived habitats throughout the 20-year span of the management interval rather than completing all treatments at one time, so that substantial patches of early successional habitat are always present.

Note in table 3 that, using even-aged management options, the overstory canopy can be very open or only partially open in newly harvested stands, depending on the practices used. As new stands grow and develop, residual trees left after harvest can influence the presence or absence of structural habitat features, and in turn, the wildlife species that utilize the developing stands. The more open the residual canopy, the more herbaceous growth, shrubs, and fruit will be produced. Exposed perches, used by forest raptors and other birds for hunting or singing, will be available in patch-cuts (small clear-cuts) and shelterwoods until they are concealed by the growing new stand. Maintaining selected overstory inclusions and large cavity trees in leave or retention patches or by careful marking of individual trees helps to build habitat structure such as large cavities in the new stand.

Hard mast availability in newly harvested patches will be low until the new stand is old enough to start producing seed again. Soft mast usually is plentiful in newly regenerating stands as long as there is minimal overstory shading. Herbaceous forage and soft mast like strawberries start to disappear in clear-cuts when the dense woody regeneration of the shrub layer shades these components out, normally in two to five years. Raspberries and blackberries stop producing soft mast when the woody regeneration grows tall enough to shade them out, in 5 to 10 years. Pin cherry grows into the overstory canopy and produces soft mast for a wide range of species including foxes, black bear, and cedar waxwings, among many others from about age 15 to 30, after which it begins to decline and die. Small hardwood and softwood "whips" — large saplings and small poles that remain after the cut — can be considered either as desirable advanced regeneration such as oak or pine and left to grow, or as undesirable (primarily other hardwoods such as beech and red maple) because their presence will inhibit oak and pine regeneration. Undesirable whips are often removed to favor regeneration of intolerant and mid-tolerant trees like aspen, paper birch, yellow birch, and white ash and so create more early successional habitat conditions. A variety of tree, shrub, and herbaceous groundcover species ensures that a variety of forage plants and insect food, nest sites, and cover elements are present.

Acorn production might be maintained over time if oaks can be regenerated. Softwood inclusions of pine and hemlock can be carefully retained or developed by releasing softwood advance regeneration. Large cavity trees normally would be present in both hardwood and softwood stands with careful retention marking, and large coarse woody debris would be present with careful marking and logging administration. If the forested parts of the property are left unmanaged due to their small size or a need for privacy, for example, one would expect little if any herbaceous groundcover and soft mast production and a closed canopy. In such cases, special efforts should be made to maintain the early successional habitats provided by old fields, hayfields, and grassy or brushy old orchards for the many early successional species that use them.

Traditional and deferred softwood shelterwood treatments create a brief herbaceous groundcover and soft mast response in treated areas, as well as a strong regeneration response of intolerant, mid-tolerant to tolerant softwood and hardwood tree species. Any oak present would periodically produce a hard mast crop every 3 to 5 years if left in the residual overstory, but fewer oaks would be found on softwood sites, so hard mast production would be low on softwood sites. Most of the mast on such sites would be seeds from cone crops. Large cavity trees would be present in both hardwood and softwood stands with careful retention marking; likewise, large coarse woody debris would be present with careful marking.

Shelterwood cutting has many of the wildlife habitat advantages of clearcutting but some very early- and late-successional species may not be present in shelterwoods. Species that only occur in very open regenerating stands, such as bluebirds and field sparrows, and species associated with closed-canopy mature stands — wood thrush, hermit thrush, and great crested flycatcher, for example, may not be present. Seeps and wet areas are found in softwood sites and are important skunk cabbage sites for black bear in the early spring and wild turkey throughout the winter. Depending on how much cull is left in the overstory, or how many cavity trees or retention areas are marked, woodpeckers and other cavity nesters may be either common or uncommon compared to the unmanaged scenario. Special efforts should be made to maintain the early-successional habitat values of adjacent open habitats such as old fields and abandoned orchards, for example.

A distribution of even-aged stands provides habitat for the most species because optimal habitat conditions for many species are always present somewhere on the property. There are distinct groups of breeding birds, for example, inhabiting regenerating, young forest, and mature forest stands. The general pattern is the same as that in all temperate forests in North America: high bird diversity in regenerating stands, low in poletimber stands, and high again in mature stands. In southern New England, no bird species need stands older than silvicultural rotation age — about 100 years. The same species occur in 80- to 100-year-old sawtimber stands as in stands twice as old. Some characteristic bird species in southern New England are: common yellowthroat, chestnut-sided warbler, cedar waxwing, and field sparrow in regenerating seedling stands; redstart and ruffed grouse in sapling stands; blue-gray gnatcatcher, ovenbird, and scarlet tanager in poletimber stands; and least flycatcher, black-throated blue warbler, and hairy woodpecker in mature stands. Some species occur in all stages from large saplings through sawtimber, for example, red-eyed vireo, blue jay, and black-capped chickadee. Others occur at forest edges, for example, Baltimore oriole and wood-pewee. Still others need a range of habitat types to nest and feed; for example, Cooper's hawks nest in mature woodlot trees but feed on a variety of small birds and mammals that are associated with woodlots and open habitats. Also, as in northern New England, ruffed grouse preferentially reside in sapling stands because they provide secure cover but they also need openings for broods to feed on insects and fruit and mature stands for winter feeding on mast and buds.

Although mammals are not as habitat specific as birds, regenerating shelterwoods and patch-cuts provide abundant summer fruit for foxes, raccoons, and black bears and winter browse for deer and moose. Regenerating and sapling stands provide secure seasonal cover for woodcock and year-round cover for New England cottontails, two species in decline throughout the region due to lack of such habitat. Mature stands provide acorns and other mast for deer, bear, squirrels, chipmunks, mice, and other species. Clear-cutting cannot be practiced everywhere in southern New England, but where possible, it is the most valuable forest wildlife habitat management practice.

Finally, the early successional stands would be cut on a short rotation, primarily in small patches. This practice is best applied in areas where there are additional upland fields, old orchards, or brushy wetlands nearby to create a "critical mass" of early-successional habitat. That is, where little open habitat exists, placing group cuts, small clear-cuts, or shelterwoods adjacent to them maximizes the amount of open habitat types for a while. Some early-successional wildlife species that might be maintained in these places include the smooth green snake, black racer, woodcock, redstart, cottontails, and red-backed vole, among many others.

Summary

In southern New England, much of the landscape contains agricultural lands, suburban developments, and forests. Other parts of the landscape are extensively forested. The management options appropriate to these two landscapes are different. In either type, however, habitat conditions in unmanaged and uneven-aged managed forests are not dramatically different, and their forest wildlife populations are fairly similar. The managed forests will have somewhat higher diversity due to early-successional and young forest conditions that are created periodically.

Where the landscape is somewhat open due to agriculture or extensive wetlands, the need to provide early-successional habitat through even-aged management is somewhat lessened. Often a substantial increase in wildlife habitat diversity can be achieved in such situations with uneven-aged management by placing group cuts near cropland, old fields, orchards, or other open habitats. Such placement increases the effective area of early successional or young forest habitats.

Where the landscape is heavily forested, uneven-aged management will increase habitat diversity somewhat by creating and maintaining the presence of components such as soft mast, young forest, and large cavity trees, for example. The greatest impact on wildlife diversity results from even-aged management, however, because a wide range of conditions from early successional habitats to old forest is present in a shifting mosaic over time.

Getting Started

AN INVENTORY CHECKLIST AND

A PRESCRIPTION KEY

IN THIS SECTION, we provide some guidance on how to get started on wildlife habitat management on your property. Often landowners hire a consulting forester to develop a management plan that can address their wildlife and timber management needs for a property (Appendix B). A management plan consists of an inventory of the property, a set of landowner goals that management activity will try to address over time, a schedule of treatments to accomplish the goals, and some monitoring of the activity to assess results.

An inventory helps you know what you have on your property. This generally starts with a map of the property. You use existing survey notes, deed descriptions, and maps to first delineate the property boundaries, roads, stone walls, and structures, fields, pastures, and so forth.

You also need to subdivide your property into forest stands—areas of forest that differ distinctly in cover type (northern hardwoods, oak–pine, etc.) and size/age of trees. You will appreciate the importance of cover type mapping as you work through the following prescription key. A consulting forester can help you delineate these stands, but it's possible to do it yourself. You can either view aerial photographs of the area at county extension offices or you can order your aerial photographs through USGS National Aerial Photography Program at <http://edcsns17.cr.usgs.gov/finder/finder_main .pl?dataset_name=NAPP>.

Wildlife habitat improvement on a forested property generally involves some timber harvesting. Unless you are equipped to do some of this yourself, you may need the services of a consulting forester who will mark the timber, arrange for the services of a qualified logger, and supervise the sale. Creating habitat patches that are large enough to attract wildlife that are not present normally involves more cutting than can be removed by residential fuelwood cutting, for example. Some noncommercial habitat treatments such as apple tree release or pruning, mast tree tending, creating or maintaining openings, or pond construction might be desirable, but you may have to spend some money to complete them.

Table 4 lists some of the important wildlife habitat elements at the larger landscape scale and the smaller within-stand scale. Getting some information on the presence or absence of these habitat elements on your own property and in the surrounding area is needed for the next step — figuring out what to do on your property to enhance habitat diversity. We use the habitat composition goals for New England landscapes from *New England wildlife: Management of forested habitats* (table 5) that provide conditions for maximum wildlife habitat diversity to guide our management activity from this point forward. You can use the following prescriptions to discuss various habitat aspects with your consulting forester. Together you and your forester can determine what types of practices you want to use to maintain or develop the wildlife habitat values on your property and promote long-term forest stewardship.

Step 1. Estimate the relative composition:

General features	My property	Surrounding area
Forest cover types:	Percent	
Hardwoods—aspen–birch, northern hardwoods		
Softwoods—white pine, eastern hemlock, spruce/fir		
Mixedwoods—oak–pine		
Nonforest cover types:		
Uplands—grass, forb, and shrub openings; pasture; savanna; and orchards		
Wetlands—sedge meadow; shallow and deep marshes, shrub swamp; bog; and small pond		
Water—lakes, streams, and rivers		

Step 2. Check for within-stand features on your property. Do you have the following wildlife habitat features on your property?

Within-stand habitat features	Lots	None	Comments
Canopy type			
Closed canopy			
Partial canopy			
Open canopy			
High exposed perches			
Exposed perches			
Hardwood inclusions in softwoods			
Softwood inclusions in hardwoods			
Large cavity trees—live or dead; hollow, nest holes present			
Hard mast—oaks, beech, hickories, walnut, hazel-nut			
Soft mast—strawberry, serviceberry, raspberry, black-berry, pin cherry			
Midstory			
Shrub layer			
Herb layer			
Coarse woody debris—large-diameter pieces			
Vernal pools			

Note: Adapted from DeGraaf et al. (1992).

Table 5

Desired Wildlife Habitat Composition Goals for New England Forested Landscapes

Composition	Habitat opportunity class	
	Essentially forest (percent)	Forest and nonforest (percent)
Habitat breadth		
Forest	>90	70–90
Nonforest	0–10	5–30
Water	<5	>5
Size-class distribution		
Regeneration	5–15	5–15
Sapling-pole	30–40	30–40
Sawtimber	40–50	40–50
Large sawtimber/old forest	<10	<10
Cover type distribution		
Deciduous (not oak)		
Short rotation[a]	5–15	5–20
Long rotation[b]	20–35	10–20
Hard mast-oak[c]	1–5	1–15
Coniferous[d]	35–50	25–50
Nonforest		
Upland openings	3–5	5–10
Wetlands	1–3	3–5

Note: [a]Short-rotation deciduous type includes the aspen-birch group; [b]long-rotation deciduous types include northern and swamp hardwoods groups; [c]hard mast includes oak–pine and oak–hickory groups; [d]coniferous type includes mixes of white/red pine–hemlock and spruce–fir groups.

Adapted from DeGraaf et al. (1992).

PRESCRIPTION KEY

In essentially forested landscapes (the subject property and the surrounding landscape is greater than 90 percent forested)	Prescription key recommendation
1 Upland nonforest and age/size class distributions meet the composition goals in table 5.	A
1' Upland nonforest component is minimal.	2
2 Age/size class distribution meets composition objectives in table 5; cover types balance hardwood:softwood types.	B
2' Age/size class distribution does not meet composition objectives; cover types not balanced between hardwood/softwood types	3
3 Where minimal disturbance, closed canopy conditions, and late-successional habitats are desired using uneven-aged management.	4
3' Where large openings, open canopy conditions, and early- and late-successional habitats are desired using even-aged management.	6
4 Low-density stand (fig. 73).	C
4' High-density stand (fig. 74).	5
5 Tree diameters mostly 10–12 inches or less (fig. 75).	D
5' Many trees over 12 inches diameter (fig. 76).	E
6 Less than 5–15 percent of the property acreage (and surrounding area) in regeneration stands up to 10 years old.	F
6' More than 5–15 percent in regeneration stands.	7
7 Remaining acreage dominated by either poletimber stands (most trees 5–10 inches diameter) or sawtimber (many trees over 12 inches diameter).	G
7' Remaining acreage has about equal proportions of poletimber or sawtimber stands.	H

In forested and nonforest landscapes (the subject property and the surrounding landscape is at least 70 percent forested with a significant aquatic habitat component)	Prescription key recommendation
1 Upland and wetland nonforest, water and age/size class distributions meet the composition goals in table 5; riparian zones present.	I, J
1' Upland and wetland nonforest, water and age/size class distribution may or may not meet the composition goals in table 5; riparian zones may or may not be present.	2

Fig. 73. Stand with well-spaced trees needing no immediate treatment. Photograph by William B. Leak.

Fig. 74. Fully stocked stand ready for harvest. Photograph by William B. Leak.

Fig. 75. Stand with small-diameter trees with limited harvest options. Photograph by Mariko Yamasaki.

2	Age/size class distribution meets composition objectives in table 5; cover-type distribution may be heavier to softwoods than hardwoods; riparian zones present.	B, I
2'	Age/size class distribution does not meet composition objectives; cover-type distribution may be heavier to softwoods than hardwoods; riparian zones may or may not be present.	3
3	Where minimal disturbance, closed canopy conditions, and late-successional habitats are desired in forest stands in more open landscapes using uneven-aged management.	4
3'	Where large openings, open canopy conditions, and early- and late-successional habitats are desired in forest stands in more heavily forested landscapes using even-aged management.	6
4	Low-density stand (fig. 73).	C
4'	High-density stand (fig. 74).	5
5	Tree diameters mostly 10–12 inches or less (fig. 75).	D
5'	Many trees over 12 inches diameter (fig. 76).	E
6	Less than 5–15 percent of the property acreage (and surrounding area) in regeneration stands up to 10 years old.	F
6'	More than 5–15 percent in regeneration stands.	7
7	Remaining acreage dominated by either poletimber stands (most trees 5–10 inches diameter) or sawtimber (many trees over 12 inches diameter).	G
7'	Remaining acreage has about equal proportions of poletimber or sawtimber stands.	H

Fig. 76. Stand with large-diameter trees would support a commercial harvest using group selection. Photograph by Kenneth R. Dudzik.

Fig. 77. Dense conifer stands provide winter cover for species like white-tailed deer. Photograph by Mariko Yamasaki.

Fig. 78. High perches serve as hunting perches for large birds of prey like this red-shouldered hawk. Photograph by Mariko Yamasaki.

PRESCRIPTION KEY RECOMMENDATIONS

A. Within balanced cover type composition, consider the following small-scale manipulations:
1. Maintain aspen–birch short rotation component.
2. Maintain quantity and juxtaposition of regenerating acres in the surrounding area.
3. Increase the hard mast component (primarily oak and beech).
4. Improve quality and juxtaposition of dense conifer thermal (winter) cover (fig. 77).
5. Continue upland nonforest opening maintenance.
6. Recognize, protect, maintain, and develop within-stand features throughout the area where possible (for example, high exposed perches [fig. 78], raptor nesting sites [fig. 79], heron rookeries, large diameter cavity/den trees, shrub layers, herbaceous vegetation, dead and down material, riparian zones, wet swales, and so on) and where needed.
7. Recognize, protect, and maintain natural special habitat such as cliffs and ledges, talus, outcrops, and caves and maintain human-made features such as agricultural fields, gravel pits, old buildings, bridge supports, and so on where necessary.

B. Consider developing the upland nonforest opening component as a part of the logging operation using cleared and seeded log landings (greater than one-third acre in size) and skid trails and roads to provide soft mast, shrub production, and herbaceous ground cover and forage. Continue with recommendation A.

C. Do nothing now. Consider recommendation A for future activities.

D. This stand is small (immature) for an uneven-aged harvest cut designed to produce new regeneration. However, there are possibilities for a partial cut (improvement cut) designed to improve the species, quality, and certain wildlife habitat values. The products that might be removed include firewood, pulpwood, some small sawlogs, etc. Pick crop trees (50 or more per acre) of valuable species with straight, clean boles and thin around them. Keep large cavity trees, hard mast trees (oak and beech), dead snags and trees that are developing into cavity and mast trees (fig. 80a and b). The partial cutting will help produce a denser understory for those species requiring this habitat condition.

E. This stand has enough large trees for an uneven-aged regeneration cut designed to harvest timber and begin regenerating the stand. Use a combination of group selection with some harvesting between the groups:

Northern hardwoods: Use groups about one-half acre in size. If the understory is mostly beech or striped maple, the group cuts should remove everything. If the understory is sugar maple or ash, cut in winter to preserve the understory. Leave important within-stand wildlife features such as large cavity trees, softwood inclusions, large coarse woody debris, and encourage shade-intolerant and mid-tolerant tree regeneration, as well as shade-tolerant species.

Oak–pine: Use groups about one-fourth acre in size, centered on groups of mature/overmature trees. Use them to release existing understories of oak or pine, **or** cut in the fall after a good oak/pine

Fig. 79. Keep an eye out for basket-shaped forks in hardwood trees that can support nests of forest raptors. Photograph by Mariko Yamasaki.

Fig. 80a and 80b. Retain trees like these whenever possible. (a) Den trees in large hardwoods are important habitat elements for a variety of mammals and birds. (b) Standing dead trees provide important nesting sites for cavity nesting birds and bats. Photographs by Mariko Yamasaki.

seed year removing all the existing understory. Leave important within-stand wildlife features such as large cavity trees and coarse woody debris, as well as maintain/improve hard mast diversity by favoring other nut-bearing species such as the hickories, hazel-nut, and the occasional walnut or butternut.

Spruce–hemlock–pine: Use small groups, about one-tenth acre in size centered on mature/over-mature trees. Release existing softwood understories if possible; if the understory is deciduous, cut it or knock it down during logging. Mark lightly between groups. Leave important within-stand wildlife features such as large cavity trees, hardwood inclusions, large coarse woody debris, and encourage intolerant and mid-tolerant tree regeneration, as well as tolerant species.

F. This property needs additional early successional acreage, up to 5 to 15 percent of the total acreage, to be achieved through even-aged management. Find stands on the property ready for even-aged harvest—stands with an abundance of large trees over 12 inches dbh and preferably larger. However, stands with an aspen component should be scheduled for clear-cutting in fall or winter to increase or maintain the aspen component. In the remaining stands, consider thinnings or improvement cuts.

Northern Hardwoods: Clearcutting of the mature stands works well. The openings should be five acres or larger. As an alternative, a deferred shelterwood may be used, leaving about 30 percent canopy closure (a very open canopy) in trees 12 to 14 inches diameter along with larger cavity trees and snags. Large clearcuts/deferred cuts may leave islands of conifers or wildlife trees. Retain large cull trees where possible to become large coarse woody debris.

Oak–pine: A standard shelterwood cut is best. Leave about 50 percent canopy during the first cut. If the understory is not oak/pine, remove as much as possible during a fall cut during a good oak/pine seed year. If there is a good component of oak/pine understory already present, try to maintain it by careful cutting during the winter. After about 10 years, remove the overstory.

Spruce–hemlock–pine: Use a standard shelterwood. Leave about 75 percent crown cover during the first cut. If there is a desirable softwood understory, preserve it by winter logging. The overstory can be removed in one or two cuts, beginning 5 to 10 years after the first cut.

G. This property, and/or the surrounding landscape, has an adequate component of regenerating stands, but a dominance of either poletimber or sawtimber. For the next 10 years, apply partial cutting to improve species and quality, and to enhance wildlife habitat conditions. After 10 years, review the situation again regarding the need for regeneration harvesting.

H. Property and/or surrounding landscape is in good condition. Partial cutting should work toward improved species and quality and wildlife habitat features. However, in 10 years, even-aged harvests will be required to maintain 5 to 15 percent of the acreage in regenerating stands.

I. Identify and manage riparian zones carefully. These zones present more habitat management opportunities for increased vertical structural diversity than do most upland stands (fig. 81). Transition zones such as these can be very important to wildlife species with significantly different habitat requirements for upland nesting and denning, and aquatic feeding locations (for example, wood duck, common merganser, bald eagle, osprey, various herons, and furbearers such as raccoon, mink,

Fig. 81. Riparian zones increase habitat and wildlife diversity in New England forests. Photograph by Richard M. DeGraaf.

and otter). Retain and develop special features such as high exposed perches and nest trees used by herons, eagles, and ospreys; large-diameter cavity and den trees; dead and down woody debris for various moist-ground-dwelling species such as redback salamanders and American toads; and areas of dense conifers for deer winter cover. Consider developing an aspen–birch component adjacent to these riparian zones. Beaver activities over time create habitat for many aquatic and early-successional upland wildlife species. Continue with recommendation A.

J. Maintain the diversity of upland nonforest cover types such as pastures, orchards, cultivated fields, old fields, and grass and forb openings periodically (fig. 82a and b). Maintain and develop the hard mast (primarily oak, beech, and hickories) component where possible. Maintain and develop at least three to five large-diameter cavity trees sufficient to provide adequate denning and nesting opportunities. Continue with recommendation A.

Fig. 82a and 82b. (a) Agricultural fields provide herbaceous forage and cover elements not found in forested stands. (b) Upland fields and openings are important foraging habitats for American kestrels. Photographs by Mariko Yamasaki.

Monitoring Your Success

As you implement your forest management plan, you will likely want to know what effect you are having. Your goal was to improve wildlife conditions — to maximize the chances of attracting more species or more of a certain species to your land. To monitor the results of your efforts, either you can make a mental note of species or sign, or you can keep a nature journal to record your observations. Recording your observations before cutting begins will give an approximate baseline for commonly occurring species.

Keeping a journal can be as simple or elaborate as your time and interests allow. If you want to track species changes you will need to be somewhat systematic. For the most part, you will be recording changes in bird species. This is so because birds are by far the most numerous species and because they generally respond to changes in forest structure more dramatically than mammals or reptiles. Keeping track of changes in the birdlife of your land means visiting the same places each year in late spring or early summer — June is best — and recording the birds singing or calling. Take your binoculars with you and use your bird guide. During the breeding season, forest birds are most closely associated with particular vegetation conditions and they sing the most. So, standing in a small clear-cut or patch cut each year in June and recording the species you see and hear in the early morning will give you a fairly clear indication of the species that use the area through time as the cut regenerates. Our memories trick us — it's hard to recall in which year we first heard a towhee as the clear-cut regenerated; a written record often surprises us when it doesn't match our recollection, so it's important to keep written notes. Another important reason for keeping records is that species may be common when you observe them but uncommon a decade or two later. Your records may be important data years later.

Similarly, winter tracks in fresh snow one to two inches deep will indicate who's been there (fig. 83). Record every track you see; photograph or sketch ones that are unfamiliar. For the most part, now you'll be recording mammal tracks. A few birds — grouse and wild turkey, primarily — will leave clearly identifiable tracks too. Winter tracking is an absorbing hobby in itself.

Besides these regular surveys to monitor your efforts to increase wildlife diversity, it's useful and fun to record individual events or uncommon sightings as you walk your land. Keep a notebook handy, and list the date, time, place, and weather conditions when you spot an owl, bobcat, moose, snapping turtle, bear, or see an unusual track. Having a camera is a real help in recording a new track or remains of a predation event; the pictures can help an experienced person tell you what was there. Field guides

Fig. 83. Bear tracks in the snow. Photograph by Jonathan Janelle.

to animals, tracks, plants, and insects are readily available and should be part of your library if you want to monitor the effects of your management.

For many, however, the most compelling reason for keeping a nature journal is the "time transport" that it provides. All woodland owners have special days, and often they involve some wildlife experience. Recording those days, with notes on the season, foliage, even the way the air smelled, allows you, years later, to recall, even relive, those days again and again — you're back in the same place, feeling the same pleasures. The value of a well-kept journal cannot be overstated.

ᗽ ᗽ

Points to Remember

- The most effective way to manage New England's forests for wildlife diversity is to periodically provide habitat patches that are distinct enough and large enough to attract species that are currently not present on the property.

- It is also important to provide stand components, such as large cavity trees, such that they will be continuously present on the property, whether using even-aged or uneven-aged management.

- Even-aged management provides habitat for more species than does uneven-aged management because it adds communities by creating large patches of early-successional habitat through time — such habitats are needed because they support a very diverse array of habitat specialist species that are declining throughout New England.

- Uneven-aged management provides habitat components, such as small patches of early successional habitat and large cavity trees, in mature stands over time.

- Even-aged management does not fragment the forest — it's still forest, with intolerant, mid-tolerant, and tolerant tree species present in all stages.

- New England's forests are neither as diverse nor as dynamic as they once were — Native American shifting agriculture and fire are no longer present, wildfire is strictly controlled, and former naturally open areas are now developed.

- Letting Nature take its course will not provide the full range of habitat conditions for all native species; given the constraints mentioned and the regional decline of modern agriculture, our forests are becoming less diverse.

🌿 Only management can provide the habitat conditions needed to maintain diverse New England wildlife populations. On most ownerships, effective management will best result when forest landowners look beyond their boundaries and consider management options that compliment habitat conditions in the surrounding landscape.

Appendix A:
Common Woodland Wildlife Species Occurring in New England

Even-aged management (EAM) creates seedling (S), sapling-pole (Sp), saw-timber (St), and large sawtimber/old forest (L) stands over time; uneven-aged management (UAM) creates stands with all size classes of trees represented within stands. Expected species presence is noted with an X. Species seasonal use of cover and foraging sites can vary greatly. See DeGraaf and Yamasaki (2001) for these details.

Species	Subregion[a]	Northern hardwood EAM				UAM	Spruce–fir EAM				UAM	Oak–pine EAM				UAM	White pine EAM				UAM
		S	Sp	St	L	U	S	Sp	St	L	U	S	Sp	St	L	U	S	Sp	St	L	U
Amphibians																					
Spotted salamander	N, S	X	X	X	X	X	X	X	X	X	X	X	X	X	X	X	X	X	X	X	X
Red-spotted newt	N, S	X	X	X	X	X	X	X	X	X	X	X	X	X	X	X	X	X	X	X	X
Northern dusky salamander	N, S	X	X	X	X	X	X	X	X	X	X	X	X	X	X	X	X	X	X	X	X
Northern redback salamander	N, S	X	X	X	X	X	X	X	X	X	X	X	X	X	X	X	X	X	X	X	X
Northern spring salamander	N, S		X	X	X	X		X	X	X	X		X	X	X	X					
Northern two-lined salamander	N, S	X	X	X	X	X	X	X	X	X	X	X	X	X	X	X	X	X	X	X	X
Eastern American toad	N, S	X	X	X	X	X	X	X	X	X	X	X	X	X	X	X	X	X	X	X	X
Fowler's toad	S	X	X	X	X	X						X	X	X	X	X	X	X	X	X	X
Northern spring peeper	N, S		X	X	X	X		X	X	X	X		X	X	X	X					
Gray tree frog	N, S	X	X	X	X	X						X	X	X	X	X					
Wood frog	N, S	X	X	X	X	X	X	X	X	X	X	X	X	X	X	X	X	X	X	X	X
Pickerel frog	N, S	X	X	X	X	X															

Species	Subregion[a]	Northern hardwood					Spruce–fir					Oak–pine					White pine				
		EAM				UAM	EAM				UAM	EAM				UAM	EAM				UAM
		S	Sp	St	L	U	S	Sp	St	L	U	S	Sp	St	L	U	S	Sp	St	L	U
Reptiles																					
Spotted turtle	S	X	X	X	X	X						X	X	X	X	X	X	X	X	X	X
Wood turtle	N, S	X	X	X	X	X						X	X	X	X	X	X	X	X	X	X
Northern brown snake	N, S	X	X	X	X	X						X	X	X	X	X	X	X	X	X	X
Northern redbelly snake	N, S		X	X	X	X							X	X	X	X		X	X	X	X
Common garter snake	N, S	X	X	X	X	X	X	X	X	X	X	X	X	X	X	X	X	X	X	X	X
Ribbon snake	N, S		X	X	X	X							X	X	X	X		X	X	X	X
Eastern hognose snake	S												X	X	X	X		X	X	X	X
Northern ringneck snake	N, S		X	X	X	X		X	X	X	X		X	X	X	X		X	X	X	X
Northern black racer	N, S	X	X	X	X	X						X	X	X	X	X	X	X	X	X	X
Eastern smooth green snake	N, S	X			X							X		X	X	X				X	X
Black rat snake	S											X	X	X	X	X					
Eastern milk snake	N, S	X	X	X	X	X						X	X	X	X	X		X	X	X	X
Birds																					
Great blue heron	N, S			X	X	X							X	X	X					X	X	X
Green heron	N, S	X	X	X	X	X						X	X	X	X						
Turkey vulture	N, S	X		X	X	X						X		X	X	X	X		X	X	X
Wood duck	N, S		X	X	X	X						X	X	X	X						
American black duck	N, S			X	X	X															
Common goldeneye	N			X	X					X	X		X	X	X					X	X
Hooded merganser	N, S			X	X	X				X	X			X	X					X	X
Common merganser	N, S			X	X					X	X			X	X					X	X
Bald eagle	N, S			X	X					X	X			X	X				X	X	X
Sharp-shinned hawk	N, S	X	X	X	X	X	X	X	X	X	X	X	X	X	X	X	X	X	X	X	X
Cooper's hawk	N, S	X	X	X	X	X	X	X	X	X	X	X	X	X	X	X	X	X	X	X	X
Northern goshawk	N, S	X	X	X	X	X	X		X	X	X	X	X	X	X	X			X	X	X
Red-shouldered hawk	N, S	X				X						X		X	X	X	X		X	X	X
Broad-winged hawk	N, S	X		X	X	X						X		X	X	X					
Red-tailed hawk	N, S	X		X	X	X						X		X	X	X			X	X	X
American kestrel	N, S	X										X							X		
Ruffed grouse	N, S	X	X	X	X	X	X					X	X	X	X	X	X	X	X	X	X
Spruce grouse	N						X	X	X	X	X										
Wild turkey	N, S	X	X	X	X	X						X	X	X	X			X	X	X	X
American woodcock	N, S	X	X	X			X					X						X			
Mourning dove	N, S	X	X	X	X	X	X					X	X	X	X	X	X	X	X	X	X
Black-billed cuckoo	N, S	X	X	X	X	X	X	X	X	X	X		X	X	X	X	X	X	X	X	X
Yellow-billed cuckoo	N, S	X	X	X	X	X						X	X	X	X	X					
Eastern screech-owl	N, S	X	X	X	X	X	X	X	X	X	X	X	X	X	X	X	X	X	X	X	X

Species	Subregion[a]	Northern hardwood					Spruce–fir					Oak–pine					White pine				
		EAM				UAM	EAM				UAM	EAM				UAM	EAM				UAM
		S	Sp	St	L	U	S	Sp	St	L	U	S	Sp	St	L	U	S	Sp	St	L	U
Great horned owl	N, S	X	X	X	X	X	X	X	X	X	X	X	X	X	X	X	X	X	X	X	X
Barred owl	N, S	X	X	X	X	X	X	X	X	X	X	X	X	X	X	X	X	X	X	X	X
Northern saw-whet owl	N, S	X	X	X	X	X	X	X	X	X	X	X	X	X	X	X	X	X	X	X	X
Whip-poor-will	N, S	X	X	X	X	X						X	X	X	X	X	X	X	X	X	X
Ruby-throated hummingbird	N, S	X	X	X	X	X	X	X	X	X	X	X	X	X	X	X	X	X	X	X	X
Red-bellied woodpecker	S												X	X	X			X	X	X	
Yellow-bellied sapsucker	N, S		X	X	X	X		X	X	X	X		X	X	X	X		X	X	X	X
Downy woodpecker	N, S	X	X	X	X	X	X	X	X	X	X	X	X	X	X	X	X	X	X	X	X
Hairy woodpecker	N, S		X	X	X	X		X	X	X	X		X	X	X	X		X	X	X	X
Black-backed woodpecker	N						X	X	X	X											
Northern flicker	N, S	X		X	X	X	X		X	X	X	X		X	X	X	X		X	X	X
Pileated woodpecker	N, S		X	X	X	X		X	X	X	X		X	X	X			X	X	X	
Olive-sided flycatcher	N						X		X												
Eastern wood-pewee	N, S		X	X	X	X		X	X	X	X		X	X	X	X		X	X	X	X
Yellow-bellied flycatcher	N						X	X	X	X	X						X	X	X	X	X
Alder flycatcher	N, S	X					X					X					X				
Willow flycatcher	N, S	X												X							
Least flycatcher	N, S		X	X	X	X		X	X	X	X		X	X	X	X					
Eastern phoebe	N, S		X	X	X				X	X			X	X	X				X	X	X
Great crested flycatcher	N, S		X	X	X				X	X	X		X	X	X				X	X	X
Eastern kingbird	N, S	X										X					X				
Yellow-throated vireo	N, S		X	X	X								X	X	X						
Blue-headed vireo	N, S		X	X	X	X		X	X	X	X		X	X	X	X		X	X	X	X
Warbling vireo	N, S		X	X	X								X	X	X						
Philadelphia vireo	N	X	X									X	X				X	X			
Red-eyed vireo	N, S	X	X	X	X	X		X	X	X	X	X	X	X	X	X		X	X	X	X
Blue jay	N, S	X	X	X	X	X	X	X	X	X	X	X	X	X	X	X	X	X	X	X	X
American crow	N, S	X		X	X	X			X	X	X	X		X	X	X	X		X	X	X
Common raven	N	X		X	X	X	X		X	X	X	X		X	X	X	X		X	X	X
Tree swallow	N, S	X					X					X					X				
Black-capped chickadee	N, S	X	X	X	X	X	X	X	X	X	X	X	X	X	X	X	X	X	X	X	X
Boreal chickadee	N						X	X	X	X											
Tufted titmouse	N, S		X	X	X	X							X	X	X	X		X	X	X	X
Red-breasted nuthatch	N, S		X	X	X	X		X	X	X	X		X	X	X	X		X	X	X	X
White-breasted nuthatch	N, S		X	X	X	X							X	X	X	X		X	X	X	X
Brown creeper	N, S			X	X	X		X	X	X	X		X	X	X	X					
Carolina wren	S											X	X	X	X	X					

Species	Subregion[a]	Northern hardwood EAM S	Sp	St	L	UAM U	Spruce–fir EAM S	Sp	St	L	UAM U	Oak–pine EAM S	Sp	St	L	UAM U	White pine EAM S	Sp	St	L	UAM U
House wren	N, S	X	X	X	X	X						X	X	X	X	X	X	X	X	X	X
Winter wren	N, S	X	X	X	X	X	X	X	X	X	X	X	X	X	X	X	X	X	X	X	X
Golden-crowned kinglet	N, S						X	X	X	X	X						X	X	X	X	X
Ruby-crowned kinglet	N, S						X	X	X	X	X						X	X	X	X	X
Blue-gray gnatcatcher	N, S			X	X	X							X	X	X	X					
Eastern bluebird	N, S	X										X							X		
Veery	N, S	X	X	X	X	X		X	X	X	X	X	X	X	X	X		X	X	X	X
Swainson's thrush	N	X	X	X	X	X	X	X	X	X	X							X	X	X	X
Hermit thrush	N, S	X	X	X	X	X	X	X	X	X	X	X	X	X	X	X	X	X	X	X	X
Wood thrush	N, S	X	X	X	X	X	X	X	X	X	X	X	X	X	X	X	X	X	X	X	X
American robin	N, S	X	X	X	X	X	X	X	X	X	X	X	X	X	X	X	X	X	X	X	X
Gray catbird	N, S	X			X							X			X		X			X	
Northern mockingbird	N, S											X									
Brown thrasher	N, S	X										X					X				
Cedar waxwing	N, S	X	X	X	X	X			X		X	X	X	X	X	X		X	X	X	X
Blue-winged warbler	S	X										X									
Golden-winged warbler	N, S	X										X									
Tennessee warbler	N	X					X					X					X				
Nashville warbler	N, S	X					X	X	X					X	X		X	X	X		
Northern parula	N			X	X	X	X	X	X	X			X	X	X						
Yellow warbler	N, S	X					X					X									X
Chestnut-sided warbler	N, S	X										X									
Magnolia warbler	N						X	X	X												
Cape May warbler	N							X	X	X	X										
Black-throated blue warbler	N, S		X	X	X	X		X	X	X	X		X	X	X	X					
Yellow-rumped warbler	N, S		X	X	X	X		X	X	X	X		X	X	X	X		X	X	X	X
Black-throated green warbler	N, S		X	X	X	X		X	X	X	X		X	X	X	X		X	X	X	X
Blackburnian warbler	N, S			X	X	X		X	X	X	X		X	X	X	X		X	X	X	X
Pine warbler	N, S											X	X	X	X	X	X	X	X	X	X
Prairie warbler	S	X										X					X				
Bay-breasted warbler	N						X	X	X	X	X										
Blackpoll warbler	N						X	X	X	X	X										
Black-and-white warbler	N, S	X	X	X	X	X	X	X	X	X	X	X	X	X	X	X	X	X	X	X	X
American redstart	N, S	X	X	X	X	X		X	X	X	X		X	X	X	X					
Ovenbird	N, S		X	X	X	X		X	X	X	X		X	X	X	X		X	X	X	X
Northern waterthrush	N, S	X	X	X	X	X	X	X	X	X	X	X	X	X	X	X					
Louisiana waterthrush	N, S											X	X	X	X	X					
Mourning warbler	N	X					X					X									

		Northern hardwood					Spruce–fir					Oak–pine					White pine				
		EAM				UAM	EAM				UAM	EAM				UAM	EAM				UAM
Species	Subregion[a]	S	Sp	St	L	U	S	Sp	St	L	U	S	Sp	St	L	U	S	Sp	St	L	U
Common yellowthroat	N, S	X			X	X	X			X	X	X			X	X	X			X	X
Hooded warbler	S	X			X						X			X							
Wilson's warbler	N						X														
Canada warbler	N, S	X	X	X	X	X	X	X	X	X	X	X	X	X	X	X	X	X	X	X	X
Scarlet tanager	N, S		X	X	X	X							X	X	X	X		X	X	X	X
Eastern towhee	N, S	X	X									X	X	X	X	X	X	X	X	X	X
American tree sparrow	N, S	X					X					X					X				
Chipping sparrow	N, S	X					X					X		X			X	X			
Field sparrow	N, S	X										X					X				
Song sparrow	N, S	X					X					X					X				
Lincoln's sparrow	N	X					X										X				
White-throated sparrow	N, S	X	X	X	X	X	X	X	X	X	X	X	X	X	X	X	X	X	X	X	X
Dark-eyed junco	N, S	X	X	X	X	X	X	X	X	X	X	X	X	X	X	X	X	X	X	X	X
Northern cardinal	N, S	X										X					X				
Rose-breasted grosbeak	N, S	X	X	X	X	X						X	X	X	X	X		X	X	X	X
Indigo bunting	N, S	X										X					X				
Common grackle	N, S	X										X	X	X	X		X	X	X	X	
Brown-headed cowbird	N, S	X	X	X	X	X	X	X	X	X	X	X	X	X	X	X	X	X	X	X	X
Orchard oriole	S											X	X	X	X	X					
Baltimore oriole	N, S		X	X	X	X						X	X	X							
Purple finch	N, S	X	X	X	X	X	X	X	X	X	X	X	X	X	X	X	X	X	X	X	X
Common redpoll	N, S	X															X	X			
Pine siskin	N, S	X	X	X	X	X	X	X	X	X	X	X	X	X	X	X	X	X	X	X	X
American goldfinch	N, S	X	X	X	X	X	X	X	X	X	X	X	X	X	X	X	X	X	X	X	X
Evening grosbeak	N, S		X	X	X	X		X	X	X	X		X	X	X	X					

Mammals

		EAM				UAM	EAM				UAM	EAM				UAM	EAM				UAM
Species	Subregion[a]	S	Sp	St	L	U	S	Sp	St	L	U	S	Sp	St	L	U	S	Sp	St	L	U
Virginia opossum	N, S	X	X	X	X	X						X	X	X	X						
Masked shrew	N, S	X					X	X	X	X	X						X	X	X	X	X
Smoky shrew	N, S		X	X	X	X	X	X	X	X	X						X	X	X	X	X
Long-tailed shrew	N	X	X	X	X	X						X	X	X	X		X	X	X	X	
Pygmy shrew	N	X	X	X	X	X	X	X	X	X	X	X	X	X	X	X	X	X	X	X	X
Northern short-tailed shrew	N, S	X	X	X	X	X	X	X	X	X	X	X	X	X	X	X	X	X	X	X	X
Hairy-tailed mole	N, S	X	X	X	X	X	X	X	X	X	X	X	X	X	X	X	X	X	X	X	X
Eastern mole	S											X		X	X						
Star-nosed mole	N, S		X	X	X	X				X	X										
Little brown myotis	N, S	X		X	X		X		X	X		X		X	X		X		X	X	
Northern long-eared bat	N, S	X		X	X		X		X	X		X		X	X		X		X	X	
Eastern small-footed myotis	N, S						X		X	X							X		X	X	

Species	Subregion[a]	Northern hardwood					Spruce–fir					Oak–pine					White pine				
		EAM				UAM	EAM				UAM	EAM				UAM	EAM				UAM
		S	Sp	St	L	U	S	Sp	St	L	U	S	Sp	St	L	U	S	Sp	St	L	U
Silver-haired bat	N, S	X		X	X		X		X	X		X		X	X		X		X	X	
Eastern pipistrelle	N, S	X		X	X		X					X		X	X		X		X	X	
Big brown bat	N, S	X		X	X		X		X	X		X		X	X		X		X	X	
Red bat	N, S	X	X	X	X		X	X	X	X		X	X	X	X		X	X	X	X	
Hoary bat	N, S	X	X	X	X		X	X	X	X		X	X	X	X		X	X	X	X	
Eastern cottontail	S	X										X					X				
New England cottontail	N, S	X	X									X	X				X				
Snowshoe hare	N, S	X	X	X	X	X	X	X	X	X	X	X	X	X	X	X	X	X	X	X	X
Eastern chipmunk	N, S	X	X	X	X	X	X	X	X	X	X	X	X	X	X	X	X	X	X	X	X
Woodchuck	N, S	X		X	X	X					X		X	X		X					
Gray squirrel	N, S			X	X	X								X	X	X			X	X	X
Red squirrel	N, S	X	X	X	X	X	X	X	X	X	X	X	X	X	X	X	X	X	X	X	X
Southern flying squirrel	N, S		X	X	X	X						X	X	X	X						
Northern flying squirrel	N, S			X	X	X			X	X	X			X	X	X			X	X	X
Beaver	N, S	X	X	X	X	X	X	X	X	X	X	X	X	X	X	X					
Deer mouse	N, S	X	X	X	X	X	X	X	X	X	X	X	X	X	X	X	X	X	X	X	X
White-footed mouse	N, S	X	X	X	X	X	X	X	X	X	X	X	X	X	X	X	X	X	X	X	X
Southern red-backed vole	N, S	X	X	X	X	X	X	X	X	X	X	X	X	X	X	X	X	X	X	X	X
Meadow vole	N, S	X					X					X					X				
Rock vole	N	X	X	X	X	X	X	X	X	X	X						X	X	X	X	X
Woodland vole	N, S	X	X	X	X	X					X	X	X	X	X						
Meadow jumping mouse	N, S	X	X				X	X				X	X				X	X			
Woodland jumping mouse	N, S	X	X	X	X	X	X	X	X	X	X	X	X	X	X	X	X	X	X	X	X
Porcupine	N, S	X	X	X	X	X	X	X	X	X	X	X	X	X	X	X	X	X	X	X	X
Coyote	N, S	X	X	X	X	X	X	X	X	X	X	X	X	X	X	X	X	X	X	X	X
Red fox	N, S	X	X	X	X	X	X	X	X	X	X	X	X	X	X	X	X	X	X	X	X
Gray fox	N, S	X	X	X	X	X	X	X	X	X	X	X	X	X	X	X					
Black bear	N, S	X	X	X	X	X	X	X	X	X	X	X	X	X	X	X					
Raccoon	N, S	X	X	X	X	X	X	X	X	X	X	X	X	X	X	X	X	X	X	X	X
American marten	N		X	X	X	X	X	X	X	X	X						X	X	X	X	X
Fisher	N, S		X	X	X	X	X	X	X	X	X		X	X	X	X	X	X	X	X	X
Ermine	N, S	X	X	X	X	X	X	X	X	X	X	X	X	X	X	X	X	X	X	X	X
Long-tailed weasel	N, S	X	X				X	X				X	X				X	X			
Mink	N, S	X	X	X	X	X	X	X	X	X	X	X	X	X	X	X	X	X	X	X	X
Striped skunk	N, S	X	X	X	X	X	X	X	X	X	X	X	X	X	X	X	X	X	X	X	X
River otter	N, S		X	X	X	X		X	X	X	X		X	X	X	X		X	X	X	X
Bobcat	N, S	X	X	X	X	X	X	X	X	X	X	X	X	X	X	X	X	X	X	X	X

Species	Subregion[a]	Northern hardwood										Spruce–fir										Oak–pine										White pine									
		EAM				UAM						EAM				UAM						EAM				UAM						EAM				UAM					
		S	Sp	St	L	U						S	Sp	St	L	U						S	Sp	St	L	U						S	Sp	St	L	U					
White-tailed deer	N, S	X	X	X	X	X						X	X	X	X	X						X	X	X	X	X						X	X	X	X	X					
Moose	N, S	X	X	X	X	X						X	X	X	X	X						X	X	X	X	X								X	X						

[a]Subregion: northern New England (N); southern New England (S). Modified from DeGraaf and Yamasaki (2001).

Appendix B:
Agencies and Organizations
that Provide Technical Information

State agencies that provide habitat management recommendations in New England

State	Responsible Agency	Address	Phone	Website
Connecticut	Dept. of Environ. Protection, Wildlife Div.	79 Elm St., Hartford, CT 06106	(203) 424–3011	http://dep.state.ct.us/burnatr/wildlife/wdhome.htm
Maine	Dept. of Inland Fisheries and Wildlife, Wildlife Resources Assessment	650 State St., Bangor, ME 04401–5654	(207) 941–4467	http://www.state.me.us/ifw/wildlife/wildlife.htm
Massachusetts	Dept. of Fisheries, Wildlife, and Environ. Law Enforcement	Field Headquarters, Westborough, MA 01581	(508) 792–7270	http://www.state.ma.us/dfwele/dpt_toc.htm
New Hampshire	Fish and Game Dept., Wildlife Div.	2 Hazen Dr., Concord, NH 03301	(603) 271–2462	http://www.wildlife.state.nh.us/Wildlife/wildlife.htm
Rhode Island	Dept. of Environ. Manage., Div. of Fish and Wildlife	Stedman Gov't Ctr., 4808 Towerhill Rd., Wakefield, RI 02879	(401) 789–3094	http://www.state.ri.us/dem/topics/wltopics.htm
Vermont	Dept. of Fish and Wildlife	103 South Main St., Waterbury, VT 05677	(802) 244–7331	http://www.vtfishandwildlife.com/

State natural heritage programs that provide information on ecosystems and rare species in New England

State	Responsible Agency	Address	Phone	Website
Connecticut	Natural Diversity Database, Dept. of Environ. Protection	Natural Resources Center, Store Level, 79 Elm St., P.O. Box 5066, Hartford, CT 06106	(203) 424–3540	http://dep.state.ct.us/cgnhs/nddb/nddb2.htm
Maine	Natural Areas Program, Office of Community Development	State House Station 130, 219 Capitol Ave., Augusta, ME 04333	(207) 624–6800	http://www.state.me.us/doc/nrimc/mnap/home.htm
Massachusetts	Natural Heritage and Endangered Species Program, Div. of Fish and Wildlife	Field Headquarters, Westborough, MA 01581	(508) 792–7270	http://www.state.ma.us/dfwele/dfw/nhesp/nhesp.htm
New Hampshire	Natural Heritage Inventory, Dept. of Res. and Econ. Development, Div. of Forest and Lands	PO Box 856, 172 Pembroke Rd., Concord, NH 03302–0856	(603) 271–3623	http://www.nhdfl.org/formgt/nhiweb/
Rhode Island	Heritage Program, Div. of Planning and Development, Dept. of Environ. Manage.	83 Park Street, Providence, RI 02903	(401) 277–2776	http://www.state.ri.us/dem/programs/bpoladm/plandev/heritage/index.htm
Vermont	Nongame and Natural Heritage Program, Dept. of Fish and Wildlife	103 South Main St., Waterbury, VT 05671–0501	(802) 241–3770	http://www.anr.state.vt.us/fw/fwhome/nnhp/index.html

State cooperative extension wildlife specialists in New England

State	Responsible Agency	Address	Phone	Website
Connecticut	Wildlife Extension Specialist	Box U87, Univ. of Connecticut, Storrs, CT 06269–4087	(203) 486–2840	http://www.canr.uconn.edu/ces/forest/dir.htm
Maine	Extension Wildlife Specialist	234 Nutting Hall, Univ. of Maine, Orono, ME 04469	(207) 581–2902	http://www.umext.maine.edu/topics/forestry.htm
Massachusetts	Director of Cooperative Extension	Univ. of Massachusetts, Stockbridge Hall, Amherst, MA 01003	(413) 545–4800	http://www.umassextension.org/topics/wildlife.html
New Hampshire	Wildlife Extension Specialist	216 Nesmith Hall, Univ. of New Hampshire, Durham, NH 03824	(603) 862–3594	http://ceinfo.unh.edu/forestry/documents/FWRhome.htm
Rhode Island	Director of Extension Service	Univ. of Rhode Island, Kingston, RI 02881	(401) 792–2474	http://www.edc.uri.edu/
Vermont	Forest Management Specialist	Aiken Center, Univ. of Vermont, Burlington, VT 05405–0088	(802) 656–3258	http://stumpage.uvm.edu/library.php

Sources of spatial information for landscape context in New England

State	Responsible Agency	Address	Phone	Website
Connecticut	Geographic Information Services, Natural Resources Center, Dept. of Environmental Protection	79 Elm St., Store Floor, Hartford, CT 06106	(203) 424–3540	http://magic.lib.uconn.edu/
Maine	Office of GIS	State House Station 125, Augusta, ME 04333–0125	(207) 287–6144	http://apollo.ogis.state.me.us/
Massachusetts	Geographic Information Systems, Executive Offices of Environmental Affairs	20 Somerset Street, 3rd Floor, Boston, MA 02108	(617) 727–5227, ext. 322	http://www.state.ma.us/mgis/massgis.htm
New Hampshire	Complex Systems Research Center, NH GRANIT	Univ. of New Hampshire, Durham, NH 03824	(603) 862–1792	http://www.granit.sr.unh.edu/
Rhode Island	Geographic Information Systems, Dept. of Administration-Planning	One Capitol Hill, Providence, RI 02908–5872	(401) 277–6483	http://www.edc.uri.edu/rigis/
Vermont	Center for Geographic Information, Inc.	206 Morrill Hall, University of Vermont, Burlington, VT 05405–0106	(802) 656–4277	http://www.vcgi.org/
All U.S. states	Electronic Data Service, Columbia University Libraries and Academic Information Systems	420 W. 118th St. 215 IAB, MC 3301, New York, NY 10027	(212) 854–6012	http://www.columbia.edu/acis/eds/outside_data/stategis.html

Timberland and woodland owners groups in New England

Organization	Address	Phone	Website
National Woodlands Owners Association	374 Maple Ave. E, Suite 310, Vienna, VA 22180	(800) 476–8733	http://www.woodlandowners.org/
Small Woodland Owners Association of Maine	PO Box 836, Augusta, ME 04332	(207) 626–0005	http://www.swoam.com
New Hampshire Timberland Owners Association	54 Portsmouth Street, Concord, NH 03301	(603) 224–9699	http://www.nhtoa.org
Vermont Woodlands Association	P.O. Box 196, 212 Main Street, Ste. 2, Poultney, VT 05764	(802) 287–4284	http://www.vermontwoodlands.org/
Vermont Family Forests	P.O. Box 254, Bristol, VT 05443	(802) 453–7728	http://www.familyforests.org/
Connecticut Forest and Park Association	16 Meriden Rd., Rockfall, CT 06841	(860) 346–2372	http://www.ctwoodlands.org/
Massachusetts Forestry Association	13 Pond Rd, Hawley, MA 01339	(413) 323–7326	
Rhode Island Forest Conservators Association	P.O. Box 53, No. Scituate, RI 02857	(401) 568–3421	http://www.rifco.org/

Organization	Website
Consulting Foresters Association of Vermont	http://www.gwriters.com/cfav.html
Association of Consulting Foresters of America	http://www.acf-foresters.com/
USDA Forest Service, State and Private Forestry, Stewardship and Land owner Assistance	http://www.na.fs.fed.us/spfo/stewardship/index.htm
MA Licensed Forester Listing	http://www.state.ma.us/dem/programs/forestry/docs/directory.doc
ME Licensed Forester Listing	http://www.state.me.us/doc/mfs/fpm/consultantlists.htm
NH Licensed Forester Listing	http://ceinfo.unh.edu/forlist.htm
MA Forestry Programs	http://www.state.ma.us/dem/programs/forestry/index.htm
VT Forestry Extension	http://stumpage.uvm.edu/
NH Forestry Extension	http://ceinfo.unh.edu/forestry/documents/FWRhome.htm
ME Forestry Extension	http://www.ume.maine.edu/~woodlot/contents.htm
CT Forestry Extension	http://www.canr.uconn.edu/ces/forest/
USDA Forest Service, State and Private Forestry, Forest Land owners Guide to Internet Resources	http://na.fs.fed.us/pubs/misc/flg/

NOTE ON THE IMAGES CREATED
TO VISUALIZE FOREST CHANGE

Landscape visualization images used in this guide were created using a series
of steps that combined real tree data and topographic information with tree
growth simulators and three-dimensional visualization programs to produce
photorealistic images of a dynamic landscape over time. These images provide
a visual depiction of forest management at a landscape level scale. The first
step involved choosing representative stand level data for northern hardwood
forest types located in northern New England and for oak–pine forest types
commonly found in southern New England. The U.S. Forest Service provided
data from New Hampshire for northern New England, which included sample plots of hardwood dominated stands, softwood dominated stands, and
vegetation from higher elevation stands. The Massachusetts Division of Fisheries and Wildlife provided data from Wildlife Management Areas (WMA)
in Massachusetts for southern New England, which included sample plots
of hardwood-dominated stands and softwood-dominated stands. The stand-
level data consisted of individual tree records containing information on tree
species, diameter at breast height (DBH) in inches, and trees per acre.

To demonstrate the dynamic nature of the stands over time, the growth,
death, and regeneration of trees were projected using a stand growth simula-
tor called NE-TWIGS. NE-TWIGS (U.S. Forest Service), an individual tree
stand growth model for the Northeast, was used to simulate changes in tree
growth and stand density over time through the Landscape Management System (LMS), version 2.0.45 (University of Washington). LMS is a computer
application that integrates several software tools that simulates forest growth
and change at stand and landscape levels. The tree data was simulated for 100
years, with output data produced every 10 years. These stand-level data represent a "snapshot" of the forest at each time step.

The next step in creating the landscape visualizations was to obtain the
topography on which the trees exist. Digital elevation model (DEM) data files
produced by the U.S. Geological Survey (USGS) consist of a grid of elevation points that were measured on the ground at regularly spaced intervals.
By using DEM files, the terrain of a particular region can be accurately created into a digital representation of the landscape. In addition to DEM files,
data files of other geographic features such as streams, lakes, and roads were
downloaded as ESRI shapefiles to produce actual locations of these features

across the landscape. ESRI shapefiles were opened and modified in ArcView 3.2a (ESRI), a geographic information system (GIS) software application that allows the user to analyze geographic information.

ArcView was also used to view the stand boundaries (represented by polygons) on the landscape as well as to assign tree data and rotations to each stand. In the attribute table of the stand shapefiles, each stand was assigned a unique stand number for identification, given a code to designate rotation, and provided a label to specify forest type. For the northern New England landscape, each stand was approximately 20 acres with one-fifth of each forest type (hardwood-dominated and softwood-dominated stands) managed every 20 years. Softwood-dominated stand management consisted of shelterwood cuts, whereas hardwood-dominated stand management consisted of a clear-cuts. Stands regenerated into a mature forest over time until age 100 when another clearcut or shelterwood removal was applied. For the southern New England landscape, each stand was approximately 7 to 8 acres, with one-fifth of each forest type (hardwood-dominated and softwood-dominated stands) to be managed every 20 years. Management of hardwood- and softwood-dominated stands was the same as for the northern New England region. This type of forest management results in at least one-fifth of the stands within a particular forest type at different stages of growth at any given time. As a group of stands matures over time, new stands in a different area of the landscape replaces their former seral stage.

The final step combines the tree data with the geographic data to produce digital images of the landscape by using Visual Nature Studio 2 (VNS2, 3DNature). Visual Nature Studio 2 is a software program that creates photorealistic digital images and animations. The DEM files and shapefiles were imported into VNS2, and "cameras" were created to produce various viewpoints of the landscape. For each forest type and age class, a VNS2 "ecosystem" was developed. Each ecosystem specified which forest species were included, how many trees there were per acre, and the height ranges of the species based on NE-TWIGS results from the LMS output. Digital photographs of real trees are assigned to each species within the VNS2 ecosystems. Since many of the tree images included with the VNS2 software were species found in the western United States with only a few of the eastern species, several tree images were modified using Photoshop 5.0 (Adobe Systems) to create trees of similar color and shape as eastern species in New England stands. Ecosystems were then linked to the shapefiles based on the forest type and rotation. Other aspects of the visual landscape images were also customized to create a more realistic illustration, such as adding atmospheric haze, cloud models, sky coloration, sun position, stream color and reflections, and road texture. Additional structures, such as power lines, buildings, and a cell phone tower, were added to the more residential, southern New England region, which also included the addition of agricultural fields. The landscape images were then rendered from different viewpoints to depict the landscape at years 0, 10, 20, and 100 and saved as bitmap images.

Links to program websites or data sources used in creating the images:

NE-TWIGS (http://www.fs.fed.us/fmsc/fvs/variants/ne.php)

LMS (http://lms.cfr.washington.edu/lmsdownload.php)

DEM from the GIS Data Deport of The GeoCommunity website (http://data.geocomm.com/dem/demdownload.html)

GIS data layers for New Hampshire by GRANIT (http://www.granit .sr.unh.edu/)

GIS data layers for Massachusetts by MassGIS (http://www.state .ma.us/mgis/massgis.htm)

ArcView software by ESRI (http://www.esri.com/software/arcgis/ arcview/index.html)

Visual Nature Studio software by 3Dnature (http://www.3dnature .com/)

Photoshop 5.0 (http://www.adobe.com/products/photoshop/main .html)

LITERATURE CITED

DeGraaf, R. M., and M. Yamasaki. 2001. *New England wildlife: Habitat, natural history and distribution*. Hanover, N.H.: University Press of New England.

DeGraaf, R. M., and M. Yamasaki. 2003. Options for managing early-successional forest and shrubland bird habitats in the northeastern United States. *Forest Ecology and Management* 185: 179–191.

DeGraaf, R. M., M. Yamasaki, W. B. Leak, and J. W. Lanier. 1992. *New England wildlife: Management of forested habitats*. Radnor, Pa.: U.S.D.A. Forest Service, Gen. Tech. Rep. NE-144.

DeGraaf, R. M. 1987. Breeding bird assemblages in managed northern hardwood forests in New England. In *Wildlife and habitats in managed landscapes*, ed. J. E. Rodiek and E. G. Bolen, 155–171. Washington, D.C.: Island Press.

Keys, James E., Jr., C. A. Carpenter, S. L. Hooks, F. G. Koenig, W. H. McNab, W. E. Russell, and M. L. Smith. 1995. *Ecological units of the eastern United States: First approximation* (map and booklet of map unit tables). Atlanta, Ga.: U.S. Department of Agriculture, Forest Service. Presentation scale 1:3,500,000; colored.

Leak, W. B., D. S. Solomon, and P. S. DeBald. 1987. *Silvicultural guide for northern hardwood types in the Northeast* (rev.). Broomall, Pa.: U.S. Department of Agriculture, Forest Service Res. Pap. NE-603.

FURTHER READING

On New England Land Use History

Bickford, C. P., ed. 2003. *Voices of the new republic: Connecticut towns 1800–1832* (Vol. I). New Haven: Connecticut Academy of Arts and Sciences.

Cronon, W. 1983. *Changes in the land: Indians, colonists, and the ecology of New England.* New York: Hill and Wang.

DeGraaf, R. M., and M. Yamasaki. 2001. *New England wildlife: Habitat, natural history, and distribution.* Hanover, N.H.: University Press of New England.

Foster, D. R. (compiler). 2002. Insights from historical geography to ecology and conservation: Lessons from the New England landscape. *Journal of Biogeography* 29(10/11):1269–1590.

Whitney, G. G. 1994. *From coastal wilderness to fruited plains: Temperate North America 1500 to the present.* Cambridge, UK: Cambridge University Press.

Wood, W. 1977. *New England's prospect*, ed. A. T. Vaughan. Amherst: University of Massachusetts Press. (Original work published 1634)

On Forestry and Forest Ecology

Beattie, M. D., C. Thompson, and L. Levine. 1993. *Working with your woodland: A land owner's guide*, revised. Hanover, N.H.: University Press of New England.

Hunter, M. L., Jr. 1990. *Wildlife, forests, and forestry.* Englewood Cliffs, N.J.: Prentice Hall.

Johnson, C. W. 1998. *The nature of Vermont.* Hanover, N.H.: University Press of New England.

McShea, W. J., and W. M. Healy, eds. 2002. *Oak forest ecosystems: Ecology and management for wildlife.* Baltimore, Md.: Johns Hopkins University Press.

Smith, D. M., B. C. Larson, M. J. Kelty, and P. W. S. Ashton. 1997. *The practice of silviculture*, 9th ed. New York: John Wiley.

On Wildlife Habitat Management

Decker, D. J., and J. W. Kelley. n.d. *Enhancement of wildlife habitat on private lands.* Ithaca, N.Y.: Cornell Cooperative Extension Service.

DeGraaf, R. M., M. Yamasaki, W. B. Leak, and J. W. Lanier. 1992. *New England wildlife: Management of forested habitats.* Radnor, Pa.: USDA Forest Service, Northeastern Forest Experiment Station. Gen. Tech. Report NE-144. 271 pp.

Gutiérrez, R. J., D. J. Decker, R. A. Howard, Jr., and J. P. Lassoie. 1987. *Managing small woodlands for wildlife.* Ithaca, N.Y.: Cornell Cooperative Extension. Information Bulletin 157.

Gullion, G. W. 1984. *Managing northern forests for wildlife.* St. Paul: Minnesota Agricultural Experiment Station. Pub. No. 13, 442. Misc. Journal series.

Hassinger, J., L. Hoffman, M. J. Puglisi, T. D. Rader, and R. G. Wingard. 1979. Woodlands and wildlife. University Park: Pennsylvania State University, College of Agriculture.

Hobson, S. S., J. S. Barclay, and S. H. Broderick. 1993. *Enhancing wildlife habitats: A practical guide for forest land owners.* Ithaca, N.Y.: Natural Resource, Agricultural, and Engineering Service. Bulletin No. 64.

Litvaitis, J. A., ed. 2003. Early-successional forests and shrubland habitats in the northeastern United States: Critical habitats dependent on disturbance. *Forest Ecology and Management* 185:1–215.

Sepik, G. F., R. B. Owen, Jr., and M. W. Coulter. 1981. *A land owner's guide to woodcock management in the Northeast.* Orono: University of Maine. Agricultural Experiment Station Misc. Report 253.

Thompson, F. R. III, R. M. DeGraaf, and M. K. Traini, eds. 2001. Conservation of woody, early successional habitats and wildlife in the eastern United States. *Wildlife Society Bulletin* 29: 407–494.

Tubbs, C. H., R. M. DeGraaf, M. Yamasaki, and W. M. Healy. 1987. *Guide to wildlife tree management in New England northern hardwoods.* Broomall, Pa.: U.S. Department of Agriculture, Forest Service, Gen. Tech. Report NE-118.

Vermont Fish and Game Dept. 1979. *A landowner's guide: Wildlife habitat management for Vermont woodlands.* Montpelier, Vt.: Agency of Environmental Conservation.

On Wildlife Natural History

Bevier, L. R., ed. 1994. *The atlas of breeding birds of Connecticut.* Hartford: State Geological and Natural History Survey of Connecticut. Bulletin 113.

DeGraaf, R. M., and M. Yamasaki. 2001. *New England wildlife: Habitat, natural history, and distribution.* Hanover, N.H.: University Press of New England.

Foss, C. R., ed. 1994. *Atlas of breeding birds in New Hampshire.* Dover, N.H.: Arcadia/Chalford.

Godin, A. J. 1977. *Wild mammals of New England.* Baltimore, Md.: Johns Hopkins University Press.

Laughlin, S. B., and D. P. Kibbe, eds. 1985. *The atlas of breeding birds of Vermont.* Hanover, N.H.: University Press of New England.

Petersen, W. R., and W. R. Meservey, eds. 2003. *Massachusetts breeding bird atlas.* Lincoln, Mass.: Massachusetts Audubon Society.

Whitaker, J. O., Jr., and W. J. Hamilton, Jr. 1998. *Mammals of the eastern United States.* Ithaca, N.Y.: Cornell University Press.

On Wildlife Sign

Elbroch, M. 2003. *Mammal tracks and sign: A guide to North American species.* Mechanicsburg, Pa.: Stackpole Books.

Elbroch, M., E. Marks, and D. C. Boretos. 2001. *Bird tracks and sign: A guide to North American species.* Mechanicsburg, Pa.: Stackpole Books.

Gibbons, D. K. 2003. *Mammal tracks and sign of the Northeast.* Hanover, N.H.: University Press of New England.

Levine, L., and M. Mitchell. n.d. *Mammal tracks — Life size tracking guide.* East Dummerston, Vt.: Heartwood Press.

Murie, O. J. 1975. *A field guide to animal tracks.* Boston: Houghton Mifflin.

On Plantings, Feeders, and Nest Boxes to Attract Wildlife

Davison, V. E. 1967. *Attracting birds: From the prairies to the Atlantic.* New York: Thomas Y. Crowell.

Decker, D. J., and J. W. Kelley. n.d. *Enhancement of wildlife habitat on private lands.* Ithaca, N.Y.: Cornell Cooperative Extension Service.

DeGraaf, R. M. 2002. *Trees, shrubs, and vines for attracting birds.* Lebanon, N.H.: University Press of New England.

Henderson, C. L. 1987. *Landscaping for wildlife.* St. Paul: Minnesota Department of Natural Resources.

Henderson, C. L. 1992. *Woodworking for wildlife: Homes for birds and mammals.* St. Paul: Minnesota Department of Natural Resources.

Terres, J. K. 1968. *Songbirds in your garden.* New York: Thomas Y. Crowell.

ABOUT THE AUTHORS

Richard M. DeGraaf, Leader of the U.S. Forest Service Wildlife Habitat Research Unit at Amherst, Massachusetts, since 1978, has conducted field research on forest wildlife in New England for more than 25 years. His primary research interest is the effect of habitat change on wildlife populations. Among his many publications are three recent books: *Trees, Shrubs, and Vines for Attracting Birds* (UPNE 2002), and two coauthored works: *New England Wildlife: Habitat, Natural History, and Distribution* (UPNE 2001) and *Conservation of Faunal Diversity in Forested Landscapes* (1996).

Mariko Yamasaki, Research Wildlife Biologist at the U.S. Forest Service in Durham, New Hampshire, has over 25 years of experience in wildlife habitat research and forest wildlife management, and is coauthor of *New England Wildlife: Management of Forested Habitats* (1992) and *New England Wildlife: Habitat, Natural History, and Distribution* (UPNE 2001).

William B. Leak, Research Silviculturist at the U.S. Forest Service in Durham, New Hampshire, has studied the effects of site and silvicultural treatment on forest development in northern New England over the last five decades, and is coauthor of *New England Wildlife: Management of Forested Habitats* (1992).

Anna M. Lester, Wildlife Biologist at the U.S. Forest Service Research Unit at Amherst, Massachusetts, has focused on developing computer models of wildlife habitat structure in upland and riparian zones. She has experience and training in stand growth simulators, geographic information systems, and three-dimensional, photorealistic visualization of forest habitat conditions.